INTERACTIONS
Stability and Organization in Biological and Artificial Intelligent Organisms

Bicky A. Marquez

Department of Physics, Engineering Physics & Astronomy
Queen's University, Kingston, Ontario, Canada

CRC Press
Taylor & Francis Group
Boca Raton London Nev

CRC Press is an imprint of the
Taylor & Francis Group, an **informa**

First edition published 2025
by CRC Press
2385 NW Executive Center Drive, Suite 320, Boca Raton FL 33431

and by CRC Press
4 Park Square, Milton Park, Abingdon, Oxon, OX14 4RN

CRC Press is an imprint of Taylor & Francis Group, LLC

Library of Congress Cataloging-in-Publication Data (applied for)

ISBN: 978-1-032-59349-4 (hbk)
ISBN: 978-1-032-59350-0 (pbk)
ISBN: 978-1-003-45431-1 (ebk)

DOI: 10.1201/9781003454311

Typeset in Times New Roman
by Prime Publishing Services

Prologue

From fundamental particles in the vacuum to the emergence of macrostructures like societies with collective intelligence, the Earth has traversed a long path that began over 4.6 billion years ago. Despite Homo Sapiens appearing on Earth only about 200,000 years ago, human intelligence has left its undoubted mark everywhere it goes on the planet and has already begun to expand throughout our solar system. After only 0.005% of Earth's lifetime, our species has pushed the limits of its adaptability via culture, technology, and innovation. Based on these principles, I wonder about fundamental questions: What allows the development of organisms with such a high level of self-sufficiency? How do the bodies of these organisms maintain a practically stable structure for so long? What keeps them consistently in one piece so that they can build and destroy alternative realities that even nature itself could not do on its own at a very fast pace?

These are questions that we will develop and debate throughout this book. Although the ultimate truth is not in our possession, we can attempt to get closer to it by discussing the theoretical frameworks within our reach and the evidence interpreted based on the most recent scientific discoveries. From the point of view of physics alone, it is challenging to understand reality if one does not delve into higher-level disciplines like biology, chemistry, anthropology, psychology, sociology, and engineering. Even though no human can specialize in all fields required to make sense of reality, we can dig into the most relevant literature supporting our exploration. There is always a risk of facing ambiguity in interdisciplinary research due to a wide range of gaps in transitions between disciplines. Nevertheless, it is a necessary framework to explain what we are today and what we can expect in the future.

From the bottom to the top, natural developments have progressively happened in a nonlinear, non-monotonous, incremental manner throughout millions of years. Based on couplings between microscopic elements, stable configurations emerged from hierarchies, synchronizations, and strong interactions that have given rise to autonomous macro-organisms. The establishment of communities, nations, institutions, social networks, and other superorganisms work together to give birth to artificial engines as a byproduct of innovation. Just as telescopes are extensions of our limited vision, or the phone elongates the reach of spatiotemporal communication between people, Artificial Intelligence (AI) will eventually become an extension of our cognitive capabilities. With the innovations in artificial cognition, we could expect to considerably amplify the scope of what we can achieve as a species. Within laws imposed by physical reality, there is no limit to what can be constructed by nature or humankind, given the resources and enough time to let combinatorics do its magic.

Contents

Billions in a Nutshell

The process of understanding complex matters is rarely linear. What we experience when we make sense of random information from the world is also hard to explain in causal terms and without ambiguities. How can we live with nuances, undefined paths, and inconclusive explanations? Humans evolved to function with a tight knowledge budget while figuring out how to navigate the world, generation after generation. With many limitations but a strong interconnection with the past through cultural infrastructures, reality slowly breaks down into more significant pieces of understanding as we advance as a species. Would these more prominent pieces of insight achieve the extension of reality at some point? Would we be able to understand reality as a whole? Or are we doomed to comfort ourselves with incrementally bigger pieces of it scattered here and there?

The intricate data stream coming our way is part of a web woven with variables primarily unknown to us. These variables could explain the random development of natural selection, the reasons behind our existence as opposed to nothing, or why some people dislike puppies. No matter how much we try to answer these fundamental questions objectively, the extension of our ambition will be limited to our anthropocentric worldview. We perceive, feel, and measure the world with our human standards, regardless of the results we get. Even if we find ways to model the behavior of nature and can predict certain aspects of it, the methods we use for this purpose are still developments from the Homo Sapiens species. Not surprisingly, what we objectively know is delimited by what is humanly possible to process for the time being.

With this disclosure on hand, let us prioritize the intricacies that make us owners of the measure of all things. The human brain is a big deal in the history of evolution on this planet. The current features of our brain represent a tipping point in contrast with other Great Apes with whom we share ancestry. The size and connectivity structure of the human brain set us apart from other hominids, laying the groundwork for mapping reality fundamentally differently than other species. The precise end-to-end mechanism through which our brains navigate the world remains hard to pinpoint. However, in recent decades, scientists have made impressive progress in this field.

Gaining insight into our complex world depends on our perceptions and ways of interacting with it. Understanding the organization and complexity of nature requires us to focus first on how our bodies interpret these mechanisms. It is fascinating to think about how a magnificent organization of neurons can emerge from complex interactions between molecular events of gene expression and environmental inputs. Trying to unveil how the brain continues to achieve higher levels of organization throughout a human lifespan is indeed intriguing[i].

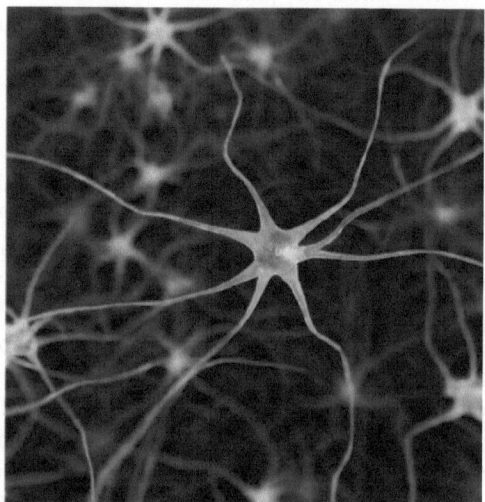

Figure 1: Neural networks.

The adult brain is composed of 84 to 100 billion neurons and approximately 100 to 500 trillion synaptic connections (Figure 1). Neurons are cells dedicated to processing information, and synapses constitute their communication medium[ii]. However, a simple description like that cannot anticipate all the complex dynamics that this extensive web of interacting neurons eventually develops over time. From this marvelous formation that we call the brain, what appears to be non-random dynamics eventually manifest. Reflecting on this, we realize that interactions between billions of similar entities lay the foundation for developing entire inner worlds characterized by uniqueness[iii].

An exciting way to understand how the brain works and emerges— still as a single entity despite the billions of elements that constitute it—leads us to dive into emergency and complexity theories. But before navigating those waters, let us start with an overview of the state-of-the-art neuroscience research supporting the upcoming discussions.

Additive Construction

A good starting point begins where highly (self-)organized systems, such as the brain, originate. The development of organized structures from fundamental particles and compounds is better understood nowadays. Nevertheless, we need to find out what caused those particles to assemble so that larger organisms could emerge. More importantly, what keeps the interactions between fundamental particles going to ensure enough stability for life to come into being? The Law of Inertia claims that an object will remain at rest unless acted upon by an external force. The alleged stability of compounded systems will persist until a perturbation enables a change. If perturbations stay below the strength of the coupling link between those particles or compounds keeping them together, the structure will adapt to the external elements and survive. Successful arrangements can admit more additions if some structural conditions are satisfied. As nothing radically disturbs this development, more giant formations may emerge and stay.

The construction of an organism has rules to follow. For instance, random elements could struggle to join the compound entity being set. For a successful component coupling to a compound (or network of elements), particular instructions must be followed—like a recipe. Otherwise, interactions between constituents may experience catastrophic consequences. In this way, when bacteria ecosystems find ways to couple with large organisms like humans, they can assist digestion and nutrient extraction from food. Probiotics have demonstrated beneficial effects on human health, enhancing quality of life. However, harmful bacteria like pathogenic strains of Escherichia coli can cause severe infections, creating strong perturbations to the stability of the host organism.

Without rules and constraints for element coupling, the stability of large organisms might be compromised. The sophistication of the rules to follow is correlated to the inner constitution of the organism that established them. The limit to the type of elements allowed to be coupled is not the sole attribute that matters. The quality of the interactions between them is critical to guarantee the stability of the structure. Living organisms have a list of instructions that contain information about their own development and stable sustainability. In a living organism, the DNA contains the necessary instructions for the proper constitution and reproduction of an organism of the same kind. Failing to follow the instructions list compressed in the DNA cannot warrant the adequate development of those life forms. From the fertilization of an egg by a sperm to birth after a proper gestation time, a long developmental process carefully curated by the intricacies of evolutionary biology is up to ensure offprints that structurally resemble their progenitors.

Stability has a meaningful role in body constitutions. It is present in the layers of material that give structure to the world we know. The stability and well-being of biological and artificial organisms are supported by interactions between their constituent elements. To illustrate the artificial case, we can imagine that exposing a computer to extremely high temperatures can lead to instability and, possibly, performance failure. Even if a tiny fraction of the billions of transistors on the CPU chip experience malfunction or failure, a general computer breakdown may happen due to a domino effect. In larger-scale organisms composed of a significantly greater number of elements, conditions for failure could be challenging to determine. As these structures are usually made up of subsystems with their own rules, large-scale entities may exhibit instability due to a chain reaction caused by multiple units failing to function parallelly.

Defining stability in complex organisms is a challenging task. At the microscopic level, an organism can look chaotic and diverse while keeping a self-sustained balance. The excitation and inhibition mechanisms behind information processing in the brain are examples of the wide diversity of compensatory dynamics that work in a coordinated manner to achieve collective purposes. Despite the plasticity that leads to crucial neural network restructuration at some stages in our lives, the brain manages to reach enough equilibrium to remain in operation without sacrificing diversity.

Representations

Around the embryonic period, millions of neurons are generated and clustered. Neurogenesis allows for forming networks that will be part of the brain's information-processing engine. Some parts of the brain, such as the Cerebral Cortex, Cerebellum, Hypothalamus, Thalamus, Pituitary gland, Pineal gland, Tonsil, Hippocampus, and Cerebral lobes, have been prestructured before birth to carry out homeostatic and other minimal functions that keep their host alive and running[iv]. These anatomical parts are building blocks that divide brain functions into areas of specialization. In particular, the Cerebellum controls our fine motor functions and sense of balance. The Hypothalamus processes information from the central nervous system and controls emotions, the secretion of hormones, and sexual behavior.

The brain structure keeps fundamentally changing until adolescence. Its dynamics is led by neuron production and consequent migration to different brain parts, following rules dictated by genetic information. The various brain regions where the migration takes place allow for the formation of mature neuron circuits. Each circuit comprises assembly material for representing reality in our heads. Representations are virtual lifeforms made from neural circuits with specific codifications and rules. The distinguishing factor among neurons lies in their unique interaction protocols facilitated by their biochemical properties and receptor sites— neurons can differ in the number of dendrites or the length of their axons. The more diversity in neuron features, the higher the chance of virtually representing complex information on those circuits.

Thanks to the diversity in functionalities and well-tuned interaction protocols between brain parts, humans can perform hybrid tasks from the start. For instance, someone who learns to draw relies on neural circuits associated with image processing, motor coordination, and gray matter while learning. These circuits are tuned to establish a specialized structure dedicated to drawing[v]. To get to this point, some clusters of neurons can get activated, and others deactivated when a task is carried out. Interactions and activation parameters allow for the representation of objects and scenarios in the brain through activity patterns.

The distribution of specialized circuits comes together as a jigsaw puzzle to represent the real world we experience. This reconstruction makes perception of reality unique for each person[vi]. The way neurons couple to each other, their activation protocol, and the resultant dynamics reach a level of diversity that is difficult to copy precisely in different organisms. Representations aim to virtually recreate the diversity of information from the constantly changing environment as accurately as possible in our minds. For this reason, virtual information is not static. As the world is independent of our perception, discrepancies between virtual and tangible manifestations of the world will eventually become apparent. To balance such a disparity, we keep modifying internal representations of our environment, correcting misleading insights about it. As a result, we each get a unique perception of everything around us that changes as we adapt to our circumstances. The more mature the representation, the more stable it becomes, allowing minor fundamental updates.

The meaning we give to incoming visual information, language, and cognition is mainly developed by circuits of neurons in the Cerebral Cortex. As cases in point, representations of perception and movement are linked to patterns formed in the Thalamus and Hypothalamus[vii]. Specialized neural circuits or networks help us embody layers of reality as accurately as possible by fine-tuning its constituent components, which includes adjusting interaction strength between neural components and specifying neural activation functions. Reaching a stable world representation is challenging since accuracy can be compromised. With our physical capabilities, we can capture a limited and defective portion of our multi-dimensional universe, where space and time vary in ways hard to understand precisely. Furthermore, the capacity to process the real-time data influx from the sensory system is narrow.

***

The interplay between reality and our internal representation of it is supported by the adaptation of individuals to a given environment. Representations are fitting models of the world that inhabit our minds. The infrastructure that helps us navigate the world emerged over generations. Nevertheless, within a generation, getting a good enough model requires feedback from the surroundings, which leads to continuous model updates until it reaches maturity. What we call the environment goes beyond the extension of physical reality. That meta-habitat includes an amalgam of situations virtually formed in our minds. The inner world of an individual, the enigmatic internal domains of those around them, and the position of everyone within a human community are parts of such an amalgam.

Adaptation enhances the survival opportunities of an organism in each environment through changes in some critical inner features over time[viii]. Many of these modifications occurred randomly over thousands of years in the evolution of species. Less fundamental adaptations that occur on significantly shorter timescales can also happen at random, but the spatiotemporal locality of the circumstances facilitates other explanations beyond the wildcard of chance.

Achieving a sufficiently accurate model of the world to navigate it autonomously can take time. Such a required maturity is not readily available for children. To guarantee their survival, parents or host communities must support them. Members of a given community might have differing maturity levels and understanding of their habitats. Still, interactions between those members should be strong enough to compensate for each other's faulty models. As part of the same environment, individuals tend to have common elements in their representations of the world and differences that make the division of labor and specialization possible. Shared features are usually customs, behaviors, ideas, institutions, beliefs, and interests encompassing the modern concept of society. Differences are based on what makes us unique and give us identities.

Regardless of the shared characteristics list, society members can modify their representations of reality to match the contours of other cultures that could even antagonize their own. Variety enriches the common understanding of the same environment for better or worse. The entanglements caused by the broad range of perceptions arise from the numerous elements that compose such a community. The reasons behind this multiplicity are complicated to enlist due to the number of components (subatomic particles, cells, bacteria, neurons, molecules) swarming around behind the scenes. Each of these constituents has the properties of autonomy, holistic indivisibility, and uniqueness. As soon as these parts interact stably to form an organism such as a human being, the predominant individual independence of each component gets compromised for the sake of the organization's well-being. However, even with this agreement, individual autonomy can still show up as the constraints that keep the organism alive are relaxed in some areas and enforced in others when necessary. The energy dedicated to keeping interconnects working is finite and might be distributed unevenly.

What amazes about stable interconnections that get to participate in the formation of organisms is that, once established, they can endure despite moderate disturbances. At higher levels of an organization, such as a community of human individuals, people connect through the interchange of perspectives, experiences, interpretations, personal information, knowledge, goods, and ideas. The reunion of all these features generates a new, more complex structure called a society. Regardless of the tensions among individual representations, society becomes an organism by strengthening common roots and agreements to establish organizational rules that may warrant stability and perdurability over time.

The New Order

Revisiting the fundamentals leads us to understand how representations are formed. Millions of interacting neurons are assembled to create structures that process information received through the body's sensory receptors. Most of these neurons are grouped into specialized circuits that help process data in various brain parts. However, these circuits are only partially isolated; they can communicate when the information has a hybrid origin—which is usually the case. To exemplify, despite visual and olfactory data being processed by different specialization areas in the brain, they can cooperate to make sense of the bigger picture.

The development of the brain's specialization areas starts from the structural rearrangements during its early stage. The number of neurons and their organization (synaptic connections and activations) varies over time. Synaptic connections and neuron generation (neurogenesis) increase and reach a substantial number in the first two or three years of a child's life[iii]. The number of neurons and interactions decreases later through pruning. This attribute comes from the fact that our brains are not ready to be used from our mother's womb. The genome information sets the ground-floor for us to explore the world, but the environment helps shape our bodies to do it correctly.

A young brain is immersed in processing loads of information absorbed through all body receptors. In this stage, a child faces the immensity of a world that embraces it with what seems to be an overwhelming number of objects and events to explore and analyze. Such a state leads to consuming a substantial magnitude of physical energy. On an average, before the age of six, children's brains are estimated to consume more than half of the total energy of their bodies (around 74%)[ix]. This incredible energy consumption is due to the need to organize a voluminous collection of neurons and synapses for survival.

Gradually, synaptic connections drop considerably during adolescence—almost by half[x]. After that period of pruning, the resultant network can be modified, but slowly and with significantly more constraints. In adulthood, the human brain consumes approximately 20–25% of the body's energy, becoming a very efficient information-processing machine[xi]. This dexterity is partly achieved by automated functionalities the brain accomplishes in many tasks. Thankfully, we do not have to plan how to open our mouths, chew, and swallow each time we eat something we have already eaten before. We also can skip thinking about how to walk on familiar routes or how to ride a bike once we have learned. Our brains automatically control all these operations.

Our capabilities are impressive because we can capture what is shared between the known and the unknown. In a new situation, we build a bridge between our current knowledge and the puzzle to understand better what is foreign to us in the new circumstance. Finding what is common to different data sources can take time and a considerable amount of energy. Instead of specifying precisely what is behind the similarities between the two datasets, we can also contrast them based on their most outstanding characteristics that can be quickly assessed. We can define these features as identification patterns in data. These patterns will help us classify the data and interact with the unknown as efficiently as possible.

The formation of a stable organization in a young brain appears to be energy-draining. It takes a certain amount of vitality to organize relevant environmental information from multiple sources to live under average conditions. Finding patterns in data contributes to such an organization. A way to find features common to various datasets is by assessing repeating tracks we identify each time a DataStream comes our way. Patterns present in data are apparent to us due to the repetition of characteristics we perceive in them. Over time, we might find certain attributes that regularly appear in different circumstances. Features that might emerge in multiple experiences and scenarios throughout our lives are defined as regularities or patterns. From repetitions, we often get information from the world. For instance, after numerous tests of throwing apples and other heavy objects into the sky, we have enough data for intuitive statistical analyses. After inspections, we can conclude that everything eventually falls if nothing stops them. We also learn that fire warms us whenever our naked skin gets close to it. The practical knowledge acquired from repetitions in data is the foundation of statistical learning[xii].

When learning to dance, we might observe the repetitive movements that performers execute and then repeat them ourselves. The pattern created by the performers through the repetition of body movements and actions is grasped and then mirrored by the apprentice until the skill is developed. These actions are reproduced in a repetitive form till the apprentice's neural networks represent the underlying motor mechanisms. Actively rerunning these steps helps strengthen synaptic connections related to the task and contributes to their durability.

***

Based on the data we can physically pull from the sensory system and the structure of our human brains, there is a fundamental limit to the type of patterns we can pick up. Beyond this human limitation, we can enable functionalities beyond our capabilities by using technology. The patterns we are directly positioned to catch are fundamental to our survival and well-being. Understanding the purpose of our biological rhythms grants the maintenance of internal stability.

Stability is afforded by keeping an internal organization in place, which implies coordinating various regulatory procedures in charge of securing the continuity of our existence. The circadian, temperature, and hormonal cycles influenced by internal and environmental circumstances are homeostatic dynamics that can be regulated to guarantee optimal body functioning[xiii]. Keeping life going is ingrained in our bodies and is optimized through experience over time. When exposed to low temperatures, we activate homeostatic mechanisms to compensate for the cold (reducing blood flow to the extremities, increment of internal heat production, and activation of hunger) and enable behavioral reactions that we learned from experience (seeking heat sources and shelters).

Insights acquired throughout varied experiences contribute to building useful knowledge. Retrospectively, we make sense of the information we collect from our interactions with the outside world. But we also employ similar operations on the internal data we receive from our bodies. Although internal and external worlds constantly change, we use repetitive data clusters (which seem almost invariant) to derive useful knowledge. The key to accessing knowledge is recognizing features that appear consistently and regularly. Data that is statistically invariant is generally observed under different conditions and measured in other contexts. Often, these invariants appear as repeated events in everyday life and are not necessarily wholly constant or ideal. However, invariants are static enough to resemble immutability.

In everyday life, humans detect patterns that become beliefs, opinions, or objective knowledge. Understanding the motion of objects, how to combine ingredients to make delicious food, or grasping the unwritten social norms to follow are examples of how we detect patterns that can benefit our lives. The quality of these beliefs, opinions, or objective knowledge can guarantee efficient decision-making and problem-solving in real-time.

The discovery of the laws of the movement of bodies has contributed to general knowledge from simple intuitions. Intuitively, people tend to deduce what is behind the behavior of objects in different situations. However, generalizing from individual experiences can lead us to wrong conclusions. For instance, after inspecting the movement of objects, someone can conclude that "objects only move if we push them, and once we stop pushing them, they will stop moving". This claim was exposed by Aristotle, who devised this proposal by analyzing the data using intuitive methodologies instead of the scientific method. A deeper inductive analysis of the evidence led Isaac Newton to conclude that a body moving at a constant speed will maintain this behavior if there is nothing to disturb it. Substantial empirical evidence and consistent data analysis are required to establish general laws that become objective knowledge. General knowledge like Newton's laws can literally take us to the moon and back.

The emergence of patterns is a precious source of certainties. Nevertheless, not all regularities we find are useful or relevant to us. The meaning of life and our position in the world could be hidden in a flux of complex data that appears to be endlessly changing. It is essential to detect clues of invariant features that can lead to certainties. Once found and proved to be such, we include them in our compilation of perpetual facts that we may categorize as knowledge. By contrast, when patterns do not emerge, disorder reigns in our heads until we find something to hook on to. Detecting regularities that can become part of general knowledge is often not a personal affair. Our peers need to assess the reliability of the data and the validity of the conclusion. If it fails, propositions will be discarded due to incompatibility with community knowledge formats.

All vs. Nothing

The brain's architecture is neither a tabula rasa (blank slate) nor an optimal organization ready to tackle the world. We do not come to the world with a completely unstructured, useless brain. Furthermore, cognitive and physical abilities to interact with the environment are yet to be tuned. Somewhere in the middle, we need to learn and shape skills for survival. The exact location of that middle point between blank slate and hardwired composition is determined by environment and genetics[xiv].

The neural architecture resulting from learning has proven to excel at capturing patterns that support our lifestyles. Just like objects thrown on a cobweb, our neural networks adhere to patterns by setting up a corresponding neural representation. Although retaining some flexibility, adult brains resist attempts that could completely modify them. It is like a sticky cobweb that cannot be easily detached from an object—if we wanted to take it off, we would have to break it or spend a long time removing each constituent part.

In the days of a child's brain formation, the reconfiguration of neural structures is much faster, and a considerable amount of energy is dedicated to brain modification-related tasks[ix]. In adulthood, these reconfigurations can conflict with the architecture in place. A fast and radical complete change in an adult brain can lead to catastrophic consequences like prolonged motor and cognitive deficits[xv]. When the organization of an adult's brain is forced to a whole new order, much of what has been learned gets lost during this restructuring process, leading to an infantile stage that emerges in an older body. By then, the maturity of human bodies does not align with the energy consumption required to restructure the brain from scratch. Without such a pump, it might take longer to adapt to the environment while the body keeps aging. A person undergoing such a modification would require assistance. Extreme cases can be seen in adults who experience severe brain damage and must make a significant effort to regain knowledge about themselves and their circumstances.

***

Neural organization and brain size are critical differences between Homo Sapiens and other species. The existence and fine development of the prefrontal cortex (Figure 2) in human beings is a crucial differentiator between our species and others, playing a significant role in adaptation to our environment. The brain's structure and development encoded in our genome are specified in the bodies of children even before they are born. A 20-year prolonged stage of formation leads to the complexity and processing capabilities that characterize our species. This interval of development takes more than 20% of the average lifetime of an individual. During this time, humans achieve cognitive maturation, accommodate their biological self (that depends on hormones, organs, and immune systems), and adjust the brain's capabilities to interact with society and other creatures in a given habitat.

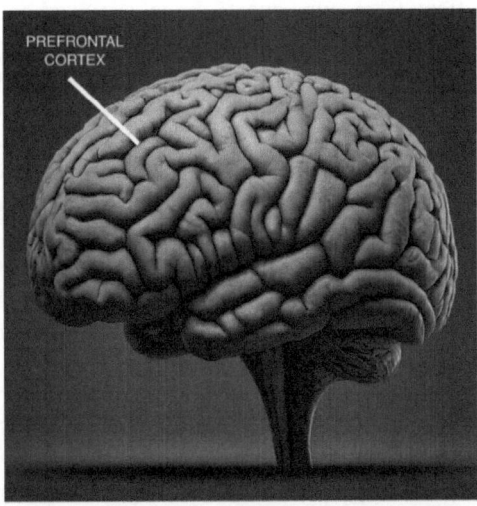

Figure 2: Prefrontal cortex.

Counterintuitively, an extensive neural network may not be the key to achieving more extraordinary skills and adaptation—a large brain is quite expensive in energy consumption and time to tune correctly. Homo Neanderthalensis (or just Neanderthals) is the species with the largest brain in the history of hominids, with an average size of around 1,500 cubic centimeters (cc). In a competitive situation against Homo Sapiens with an average volume of approximately 1,350 cc, a long-term maturity stage of the brain could have delayed the development of essential skills that hampered the survival of Neanderthals[xvi]. Greater complexity enabled by a larger brain can be beneficial in the long term if the carriers are not menaced by their surroundings and have the conditions to get there safely. In contrast, brain size is not alone correlated with great cognitive skills. Elephants' brains are ~ four times larger than Sapiens, which may not correspond to higher cognitive abilities given the proportion to their body size. Additionally, the average weight of an adult elephant's brain is only 0.1% of its body's weight, while a human's brain is about 2%.

Furthermore, the proportion is not everything since the Etruscan Shrew (Suncun Etruscans) has the most significant known brain-body size ratio with a percentage of 10 without achieving human's high cognitive capabilities[xvii]. Despite the unfairness behind comparing human cognitive abilities with others in an anthropocentric manner, we do it often. From these analyses, we hypothesize that the Suncun Etruscan's intelligence differs from ours due to brain organization specifically developed to tackle challenges in its ecological niche.

A prolonged brain formation is also one of the many requirements to achieve the high level of complexity involved in human-like intelligence. Salamanders have a slow rate of brain development, just like other amphibians[xviii]. Unlike mammals, these organisms have different priorities, including body development first. Yet, Salamander's body mass and brain constitution still differ from their human counterpart. As far as we can tell, we can find other species with higher body mass ratios, longer development times, and similar DNAs and brain structures. However, what differentiates human brains (like the size of the cortex) from others creates a significant divergence.

Despite all the wonderful aspects of the constitution of our bodies, a great gift is often accompanied by a curse. The complex way human brains are pre-organized before birth makes it much slower for them to develop than other species. At an early age, the generation of an excess of synaptic connections dedicated to information processing is expected. During this period, children must learn how to classify, filter, and redistribute all the data entering their brain regions (Limbic system, Brain stem, motor and somatosensory cortex, lobes). A pruning stage follows this step to eliminate what is not used. A long process of this kind corresponds to inefficient brain development. Human children are less able than other species to fend for themselves. In contrast, Robin birds can fly two weeks after birth, healthy foals can walk after two hours, while human babies learn to walk and communicate through language over many months according to our species' time scale.

After a particular brain organization is achieved and the human being is adapted to its environment, catastrophic instabilities that can negatively impact them still can occur. Some devastating diseases, such as Alzheimer's or dementia, represent extreme brain disorganization. This condition is undesirable since Alzheimer's can destroy meaningful synaptic connections in the brain and cause neural cell death, leading to decreased cognitive functions and reduced brain volume[xix]. The result is a different form of brain organization that could be more understandable to other community members and less useful for survival in their habitats. Brain structures radically dissimilar to what is considered average by a given community and environment can be interpreted as disorganized or even unstructured. These new forms of neural organization might be so different from the average that it becomes hard to incorporate them. Therefore, we tend to define them as disorganized.

Development

As complex organisms, our constitution is determined by layers and layers of interacting elements. The assortment of fundamental particles and the strength of their interactions assemble the universe we know piece-by-piece. Strong interactions between fundamental particles allow the existence of atoms as identifiable entities. Atoms are composed of protons, neutrons, and electrons. Protons and electrons have positive and negative charges, respectively. As the charges have opposite signs (positive and negative), electromagnetic forces lead to the attractive behavior between these particles.

What keeps neutrons and protons bound together in the atom's nucleus relates to a process different from electromagnetism. Due to the neutron's neutral charge (no charge), electromagnetic fields do not help to keep it bound to the proton in the atom's nucleus. Strong forces between protons and neutrons are established to hold them together in the nucleus (Figure 3). These forces can be interpreted as communication channels between both particles. In this process, the information traded between particles happens by exchanging a fundamental particle called the gluon. Gluons are the mediators of the strong interactions that keep elements in the nucleus of an atom together.

According to the Quantum Chromodynamics theory, the strength of interactions between particles depends on their relative distance. The shorter the distance, the stronger the force that binds them together—and the sturdier the information exchange is. Since the distance is miniscule, the exchange of information between particles happens rapidly since the data does not need to travel long distances between the transmitter and receiver. The time scale at which these exchanges happen is so short that it appears instantaneous to any organic sensory system. If a fraction of a second is complex to acknowledge, imagine how challenging it would be to perceive an event happening in less than a billionth of a trillionth of a second[xx].

Figure 3: Atom: We find neutrons and protons inside the sphere, and
the orbits correspond to electrons.

The stability of an atom nucleus is therefore supported by small
particles near each other, interacting at an ultrashort time scale. These
properties are robust enough to assemble a stable compound like the
nucleus. From the perspective of an observer with a larger spatiotemporal
scale, an atom behaves as a single unit. In this way, the identity of the
nucleus of an atom is formed out of high-speed strong interactions that
keep protons and neutrons practically hard to separate.

We are several electrons away from the description of nuclei to
atoms. A less strong (electromagnetic) force maintains electrons orbiting
around the nuclei of atoms. The attractive nature of particles with opposite
charges ensures that a nucleus (positive charge) and electrons (negative
charge) are bound. What is behind the existence of an electromagnetic
field has yet to be determined. Some theoretical frameworks predict the
existence of a virtual photon that enables the exchange of information
between electrons and the nucleus. While the speed at which this
exchange happens is yet to be defined in the conventional sense, we can
imagine it to be instant to the naked eye. Fast interactions may warrant
the necessary stability that grants an atom its identity. Such stability is
supported by particles in the atom staying together.

Ever since the Greek philosophers of nature, Leucippus and
Democritus, theorized about indivisible particles called *Atomos*

('indivisible' in Greek), the scientific community has been developing a variegated number of theories that aim to explain how matter is composed. Now we know that atoms are not indivisible (we can divide them into protons, neutrons, and electrons). Still, to our knowledge, we found other entities, such as quarks (which compose protons and neutrons), that are candidates to be defined as such since they are indivisible. Building up layers of complexity, from a single quark to the formation of atoms containing multiple quarks, we have already increased the number of elements interacting in a stable enough condition[xxi].

The emergence of more complex structures does not cancel the distinctive identity of their constituting parts. In this way, molecule formation relies on clustering atoms that share their electron clouds without risking losing the properties that make atoms what they are. As such, a water molecule is constructed based on the interactions between two atoms—one hydrogen and one oxygen. If a hydrogen atom has one electron, and an oxygen atom has eight electrons, therefore a water compound has ten electrons in total. Four of these ten electrons will group in pairs to make covalent bonds. With a pair of electrons from different atoms bonding to achieve a stable configuration, water molecules appear to us as identifiable entities.

Interestingly, the more the elements are shared, the stronger the bonds can become. A triple bond is formed when two atoms share three pairs of electrons. Triple bonds between identical atoms are shorter than double bonds, and since breaking all three bonds demands more energy than breaking two, a triple bond is consequently sturdier than a double bond. Despite the robustness of the molecule with the most bonds, environmental conditions such as high temperature can make pairs of electrons in a covalent bond split. High temperatures are associated with more energetic movement of molecules, making them susceptible to breaking. For instance, at very high temperatures (above 2000° C), water molecules can decompose into hydrogen and oxygen atoms. Likewise, higher energy environmental agitations may decay atoms into more fundamental particles. Like everything else, things do not exist in isolation. Even the most stable configurations that ever existed are prone to be disturbed by their environment[xxii].

The survival of a complex structure depends on the quality of its internal composition and the good behavior of its circumstances. The environment establishes the conditions for these compounds to form and remain. However, as all actions have consequences, newly created configurations may also impact the conditions of the environment, leading to a cycle of regulation and disturbance among all the involved parties.

***

Despite all the efforts to guarantee highly stable architectures, many different conditions can lead to catastrophic disruptions. Extensive systems eventually emerge with more elements with a spectrum from weak to strong interaction strengths. Some objects formed by a high diversity of interacting elements may be more prone to be thrown out of their natural balance even by a weak turbulence. Whatever external disturbances that cause a catastrophic change in a large system may lead to unpredictable structural behavior. Some other compositions may be more robust and require strong disturbances to be taken off balance. What is behind the origin of stability is a subject of intense study up to our days.

There are reasons to believe that some primordial soups (or seas) were linked to the beginning of life on Earth. The characteristics of such hot seas are similar to what underwater volcanic vents can generate. The conditions down there resemble a natural laboratory where many molecules already present in our organisms were mixed with high temperatures and chemically enriched water (Figure 4). How the transition between biological experiments in the primordial laboratories and the origin of life occurred still needs to be clarified[xxiii]. However, much has been speculated about the precursors to modern cells coming from those seas. In this swarm of possibilities, the so-called protocells are at the forefront of the prebiotic models that attempt to explain the transition between non-living and living matter[xxiv].

The emergence and decline of so many objects constituted by the concatenation of various elements do not happen out of anywhere. Changes in the environment play a dramatic role in the mixing of components. After all, the ingredients in a soup can be slightly mixed in the absence of high temperature. Still, a good soup recipe needs around one hour to be cooked over medium heat to transform into a delicious blend. For this reason, Paleoanthropology research is developed in conjunction with the study of climate change, biology, and geology to explain the origin of life.

No matter how many scenarios we can entertain, we have yet to truly understand the origin of life on Earth beyond the range of mythological

Figure 4: Beginning of life on Earth.

and religious explanations. Defining what is alive and what is not is complicated. If we define an object or event and its opposite, we should describe what is common and uncommon to both. When a natural entity we want to elucidate is reduced to an abstract model disconnected from everything around it, chances are that we find ourselves constantly adjusting the model to incorporate a practical context into it. For instance, a circle is defined as an arrangement of points in a plane that are equidistant from a point in the center. However, a more nuanced definition would consider asymmetries caused by random points unequally distant from the center reference point. This extended description will be followed by an assortment of explanations (context) required to accommodate the existence of those asymmetries. Likewise, when defining what is alive or not, researchers need help to draw the line when considering the richness of complex micro-entities that inhabit our planet.

Reproduction, adaptation, homeostasis, growth, genetic information, responsiveness to the environment, and metabolism are among a group of features that, one way or another, a living cell has. The absence of these attributes will make for a non-living being. The interactions between those characteristics play a fundamental role in what is considered alive. In this debate, we find creatures like viruses that may suffer from the Schrodinger cat dilemma; they may be dead or alive, but not simultaneously. These discussions are relevant because, depending on the model we adopt to define something, information can be discarded or added to the entity (alive) and its opposite (non-alive)[xxv].

With the above reflections in mind, we can see how finding the correct elements to build a protocell model that can serve as a non-living precursor of living cells has been challenging[xxvi]. Without any designer who can decide how an organism will be structured or what its functions should be, a protocell would have emerged out of a self-organized process. There, interactions between the environment and the constituting materials would have been involved in an incommensurable number of combinatoric trials at random to organize an arrangement of molecules into a perdurable entity.

About perdurability, it is essential to remark that time is hard on us all, and we cannot escape its eventual consequences. The strong force that attracts protons and neutrons in a nucleus together can act in the vacuum at practically any given time. Some random atoms can bind together to form inconsequential molecules under some conditions and stay detached under others. Going up in the scale of larger configurations, a stable network of molecules would be even more challenging to form due to all the subjacent elements and conditions that need to be in the right time and place. If all these processes are a product of random events in a confined space like a primordial sea, what enabled perdurability out of pure randomness? Suppose the conditions for emerging stable complex molecular compounds are particular and hard to replicate. In that case, such compounds must have some non-random replicable mechanism to guarantee its continuity over time.

Say that the primordial soup hypothesis is correct. With the right ingredients, temperature, and other requirements, the origin of the first cells could have been found in these soups. In a given soup, mixing lipid-like molecules, high temperatures, chemical reactions, etc., would have given birth to cells. However, soups like that may not have been able to generate all the cells that exist up to the present time. Out of that specific circumstance, a replicating mechanism would be required to enable their existence on Earth. Replication is not an isolated process; it needs access to energy sources and survival in environments with beneficial and harmful substances.

Constraints

It is interesting to wonder about the stipulations under which a molecular compound cannot grow infinitely in size; instead, it forms simple structures like cells. Among many features that can limit the infinite development or growth of an object, a remarkable highlight in the field of emergence is the existence of constraints. Constraints are features that can guide the development of specific dynamics without active decision-making on how the process should unfold. Back 4 billion years ago, when the first cells' genesis occurred, the confinement of prebiotic elements and the type of available chemical reactions in a delimited space were definitive. Such confinement put into place the constraints that enabled the combinatorics game to perform its magic.

In a non-constrained situation, the number of elements and possible ways of recombining them can be infinite, leaving this game to pure randomness. However, when constraints are added to the model, the number of possible combinations is limited, removing the white noise type of randomness that could otherwise define the ecosystem. When white noise rules a system, patterns may appear but not persist over time. The events that make up this type of noise are independent. This independence cannot support perdurability as events will not repeat consistently since the space of possibilities has no limits and no way to return to a previous state. Following this logical trend, when sheep are confined by the guidance of a Shepherd dog herding them, we can see how the resultant sheep's group dynamics unfold in time. The sheep group together creating amazing geometrical patterns that are the opposite of random. Such an organization is hard to achieve in a finite time without some constraints. Constraints are what make organization possible.

Without leaving everything to pure randomness, the confinement of the prebiotic elements and the transformative properties of chemical reactions allowed the generation of new compounds. Chemical reactions can break the chemical bonds that give molecules their identities. Conversely, chemical bonds can also be formed with the right conditions. In this way, water is created when hydrogen gas reacts with oxygen gas at 300° C. The variations in temperature back in the primordial sea might have enabled a large diversity of possible new compounds and reactions over time, which led to the development of organic material from inorganic sources.

In 1828, Friedrich Wohler synthesized organic molecules in the laboratory using inorganic compounds only[xxvii]. The synthesis of urea (the crystalline substance present in the urine of mammals) happened by gently heating silver cyanate (which contains one carbon atom) with ammonium chloride in an isolated container. This accidental discovery was a breakthrough because, for the first time, organic material based on inorganic elements could be developed in the laboratory. This fact shows us how many reactions and combinations of elements in constrained circumstances could have enabled the development of living organisms primarily composed of organic material. Urea is considered an organic compound because it contains carbon and is also found in many living organisms.

Life is said to be based on carbon. Why is that the case? The answer might take us back to the core feature described in the first chapter of this book: stability. Stability is an essential property of compositions made from more than one element. Without the stability provided by the base layer of any complex architecture, survival is not guaranteed in the long term. The secret behind the construction of a tall building is the quality of its foundation. We can confidently state that stability is a fundamental attribute of enduring complex structures.

As suspected, carbon is fundamental to life because it forms stable bonds with many elements like oxygen, nitrogen, or hydrogen. When carbon is a frequent component in vast and complex structures, their subsistence is ensured, at least from the foundations. Thankfully, our bodies' organic compounds are carbon-based—DNA, lipids, proteins, etc. Importantly, it is to highlight that the origin of the stability of our bodies comes from material formed inside the boiling core of stars. Stars expel this carbon enriched material into space in the shape of stardust or large chunks of rubbish that could contribute to creating planets—this is how elements like carbon ended up in primordial seas[xxviii].

On Repetition

Until now, we have exposed the need for some constraints for developing consistent dynamics. Defining limits to what can be done in a given circumstance with finite resources comes with some penalties. It allows for combining and transforming those elements in different ways, but it also provides room for the appearance of repetition. Repetition permits the existence of cycles necessary for consistency and replication. When there is a process to be followed to form an object, there is usually a sequence of rules behind the scenes that enable the duplication of that object all over again. The cyclic nature of repetition and reproduction can be defined as recursivity, an inherent quality of life.

The recursive attribute of simple systems such as fractals shows how objects can be reproduced from simple iterative rules. These geometric shapes appear in nature as never-ending patterns with self-similarity.

For more complex structures such as living systems, self-similarity in the reproduction of patterns might not be as apparent to us. However, some recent studies point in that direction. In 2015, researchers from The University of Tokyo found that recursion may play a fundamental role in the ability of protocells to self-proliferate for multiple generations[xxix]. In this model, the authors describe a primitive model cell (protocell) that demonstrates the various cyclic phases present in the development of its persistent existence. In such a model, the cycles of ingestion, replication, maturity, and division can be selectively activated by circumstantial external stimuli, such as the thermal cycle, to produce more protocells in a process that attempts to mimic cell division.

Even though this study does not explain the complexity behind the cell division of living organisms, it is a stepping stone to understanding these intriguing phenomena. Models comprising cycles of four discrete phases (ingestion, replication, maturity, and division) for cell division might be too simple to explain life (Figure 5). Still, it shows how the self-organization of elements in primitive conditions could support explaining what remains behind the perdurability and evolution of living organisms.

Suppose, the process of recursion sheds light on how primitive cell division is possible. In that case, there might be a link between the protocell model described in that study and the development of fractals in nature at a larger scale. It is important to remark that protocells are not actual cells; these hypothetical spherical collections of lipids are precursors to modern cells only in theory. Finding the link between protocells, prokaryotes, and eukaryotes' cells is crucial to understanding the origin of life. The primitive reproductive mechanism of cell division

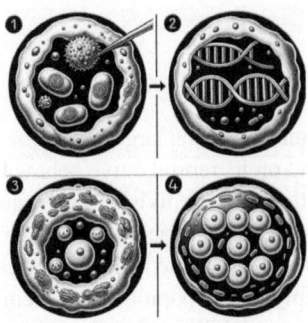

Figure 5: Cycles of ingestion, replication, maturity, and division.

in living organisms must be reproduced in the laboratory before we can claim to understand it.

Beyond cell division as a reproduction process in simple organisms, large multicellular organisms like mammals reproduce sexually in wildlife, meaning that a fusion is required between reproductive cells from opposite-sex entities to form an offprint. This coupling mechanism between two complex entities (male and female) to create new organisms is also found at many other different levels in nature. As we will see in the forthcoming chapters, coupling is behind the emergence of large systems in biological and artificial organisms. Coupling dynamics not only enable sexual reproduction in hominids, but it can also facilitate the emergence of dynamics like simple synchronization between clocks or globalization of information and trade between foreign communities.

Apart from sexual reproduction found in large organisms like mammals, other breeding methods exist across different organisms. For instance, some birds, insects, and amphibians can produce offspring from eggs without fertilization[xxx]. This virgin-birth form of asexual reproduction is called parthenogenesis and is considered routine in invertebrates such as ants, wasps, bees, scorpions, etc. One of the rare cases was the asexual reproduction of a female California condor. In this case, a cell in a female exhibits a behavior that can be compared to a sperm when merging with an egg. This typically happens in animal groups where breeding males are scarce or absent. In the laboratory, scientists can also create human embryos without eggs or sperm. These experiments are hard to run entirely due to the solid ethical and legal barriers involved. The limits of what can be done will be dictated by nature. As long as the laws of nature permit, reproduction in the wild or the laboratory will be possible despite artificial laws implemented by a given society.

***

Due to the finite nature of living organisms, the survival of a species will ultimately depend on its reproductive behaviors. Even for primitive cells, some reproduction via cell division was required for their survival and, later, for the maintenance of large multicellular organisms such as Homo Sapiens. Multiple cell-based organisms cannot exist without a mechanism for multiplying the population of elements.

In addition, the availability of nutrients, specific environmental conditions, internal well-being, age, the accessibility of mates (if required), and the social position in their community influence the proliferation of offspring. For humans, these variables get to a very different scale compared to cells like some prokaryotes that may only need carbon sources, a pH of 7, and between 20° C and 40° C to grow[xxxi]. The list of requirements for successful individual survival and the multiplication of the human population is considerably more extensive. For that level of elaborateness, a sophisticated administrator like the brain coevolved to support development and survival in different ways. Brains evolved for thousands of years when groups of neurons started communicating more closely, creating the central nervous system. More recently, in the lengthy scale of evolution, brains for our hominid ancestors emerged around 7 million years ago. *Sahelanthropus tchadensis* (nicknamed Toumai) was an inhabitant of the Sahara, found to be the earliest of the extinct species in the human family so far[xxxii]. The brain size of this species (350 cc) was only around 26% of the modern human brain size (1,350 cc). Despite Toumai's cranial capacity being closer to a modern chimpanzee's skull, this creature from northern Chad has features suggesting a mix between humans and apes.

The intricacies of keeping macro-organisms alive at the mammalian scale are enormous. We can think about all internal organs and processes that carry out vital functions for life. Each organ is a full-on system working to create, transform, or discard elements in the body. The heart, lungs, liver, stomach, small and large intestines, ears, eyes, endocrine glands, pancreas, bladder, uterus/testes, skin, spleen, and kidneys are there performing essential functions and regulations semi-autonomously—in contact with the nervous and endocrine systems, but without conscious intervention from the brain. Every organ is, therefore, a structure based on cells organized in a way that warrants stability under certain environmental conditions.

To compare, fundamental particles (that compose our atoms) interact in less than a billionth of a trillionth of a second (zeptosecond), and cells interact at a considerably larger time scale, but are still invisible to the human's naked eye. Neurons (nerve cells) exchange information through synapses in the order of a thousandth of a second (milliseconds). Other cell types have different exchange rates. All these ultrafast interactions between elements at tiny spatial scales keep us in one piece. The information exchange rate between elements is essential in forming biological or artificial organisms.

The brain size of Toumai had likely served the purpose of sustaining its life, but it might not have been adequate for a human body. The evolution of the human brain since the appearance of the known first hominids dates from 7 million years ago. Only 200,000 years ago, the cranial capacity converged to 1,350 cc. This large brain capacity evolved to include an expanded neocortex, which occupies about 76 to 80% of the human brain weight. The neocortex involves sensory awareness, spatial cognition, conscious thought, movement directives, event recall and semantic memories, language, social and emotional processing, and learning processes. These highly developed functions are the latest additions to the brain in evolutionary terms. Even though other mammals have cortexes, the human neocortex is considerably larger and more complex than other species[xxxiii].

The high-level functions of human brains were not only the product of keeping a complex organism alive and stable but also of coordinating with other members of the species, which also happened to be highly complex themselves. The development of social skills was crucial for maintaining a proper balance. Social interactions are not simple bidirectional exchanges of particles between neutrons and protons or signals between neurons. Social interactions have many shapes and forms resulting from the combination of internal processes related to high functions in the individual brains, together with different communication venues modulated by the sensory system.

The assortment of communication media and collective information might have given birth to what we define as culture. The speed of data exchange between people differs significantly from subatomic particles or neurons. For example, understanding what someone else is trying to communicate can have diverse scenarios depending on the chosen medium. If two persons speak the same language and know the exact word repertoire, simple communication through spoken language might take hundreds to thousands of seconds.

Exchanging information at the macroscopic/human scale might take longer. The strength and stability of the involved interactions might be affected by the rate at which they happen. The efficiency level at which communication unfolds has many variables and perspectives. In the next chapter, we will discuss how communication coexists with the development of our species.

Communication

Communication is all about interconnects. From the micro-cosmos to the macro-world, the exchange of information between elements of the same family has been essential to the emergence of large inanimate entities and living organisms. Information transfer can be rated in terms of its delivery efficiency. Skills required to reach highly efficient communication ranks in complex macro-organisms like humans must be learned.

At an early phase, children accumulate abundant synaptic connections. Many of the high cognitive functions, such as language, and interpretation of sensory information in children begin to develop rapidly in this early period. The number of interactions between the outside world and the toddler increases, with little substantial filters that could hinder these interactions. A child tastes what is around them through sensory communication with the outside. From these experiences, a database forms with data collected by the sensory system. Such a database is a multidimensional version of the world, and it is organized to support the fine-tuning of children's capabilities over time. Tuning redesigns a neural network multiple times until a functional arrangement emerges. After this, updates to the neural architecture are still possible but will be less fundamental. Most of these connections may disappear or change sometime later. In this last stage, neurons and their infrequently used interconnects may disappear[xxxiv]. Such a pruning process can be seen as a natural selection for neurons.

With substantial interaction with the elements of its environment but scarce high-level filters, a receptive brain could try to absorb more than it can process. Since perceptual narrowing has not been established, brain tuning depends mainly on experiences in the host environments[xxxv]. At this stage, there is a lack of discrimination about what requires priority attention in a given environment.

The dataset with which a human brain is trained takes advantage of scientific, technological, historical, artistic, and cultural knowledge.

Values, beliefs, understanding, skills, practices, experiences, observations, and other cultural features are critical to interpersonal relationships and navigating the world more efficiently. The guidance provided by information exchange through solid interactions with others may empower individuals toward a faster and better understanding of reality.

Culture is a crucial factor in understanding how knowledge is constructed. Beliefs and norms in communities endorse directions when pursuing insights about our existence and position in the universe. Culture is "the group-typical behaviors shared by community members that rely on socially learned and transmitted information"[xxxvi]. Skills, beliefs, practices, and artifacts are transmitted to new members of a society[xxxvii]. This type of interaction leads to learning, where education is supported by cultural transmission and established filters posed on sensory data[xxxviii]. According to the precepts at the time, the information our communities have accumulated for centuries has helped humanity carry out brain training as efficiently as possible. In this way, new generations do not need to reinvent the wheel but keep building on top of knowledge. A world in which information is not transmitted from generation to generation would not have enabled our species to enrich knowledge as efficiently as we do. Perhaps, Homo Sapiens would be partly like any other species with insufficient worldwide inter-generational knowledge to promote large-scale technological, scientific, and social developments.

Since we have knowledge passed down from generation to generation, we might wonder how we cannot teach all our expertise to newborns. A partial answer can be found in that our neural networks are, in part, trained to support interactions with our surroundings. Unused resources are lost over time, as with children who learn and forget multiple languages over the years. Foreseeing what the most relevant tools and knowledge for the development of an individual will be without detailed context is not easy[xxxix].

Using this reasoning as a foundation, we can consider the worth of expecting a future society with technology to load information into our brains directly—just like we would see in the science fiction movie, *The Matrix* (1999). Suppose, we get to the point where we can understand the brain adequately enough to offer accelerated learning services through specialized devices. In that case, we could modify our neural networks to perform complex tasks within a few hours or days. This scenario could become a reality if an interface with efficient, biologically realistic neural network training methods can be designed. A resource like this looks limited in the long term by the possibility of complete forgetfulness if the brain prioritizes other things.

If forgetfulness is not achieved when needed, the brain could lose the required plasticity to adapt to its ever-changing environment. After all, the human body is finite and evolves with each breath of life, so the brain must react to environmental variations and tune itself to continue interacting. A brain that does not change at all will not be able to function appropriately for the purposes for which it exists.

***

A trained mind can build complex thoughts, plan actions, and predict consequences—all relevant to the circumstances. We can interpret prediction in this context as the ability to estimate the effects and consequences of actions to be taken. To make accurate predictions, the brain must be specialized in the field of interest. A table tennis player needs to be able to make real-time predictions of the future trajectory of a moving ball. The trained player should anticipate when the body must perform the necessary movements to hit the ball successfully. Repeating the same activity with minor variations will help strengthen synaptic connections.

Learning skills generally make us better at processing real-time data, prompting faster pattern detection and enhanced decision-making. Anticipation leads to more efficient decision-making as informed decisions are made quickly enough based on experience. The more experience in a specialized field, the lower the uncertainty in deciding the next step. The difference between a veteran painter and an apprentice is found in the ability of the former to assess the right line angles that will compose a portrait or the color gradients necessary to implement depth. Eventually, both the master and apprentice might be able to reproduce the same outcome. However, depending on the subject's complexity and the skills, the apprendice will achieve it later.

The part of the human body that enables anticipation is lodged in the Cerebellum and the Basal Ganglia[xl]. In particular, the cerebellum plays an essential role as its volume increases when subjected to continuous training. Accumulating information for brain training is an individual task and part of a social phenomenon. Information shared among peers and generations supports the development of communities and the education of individuals based on the propagation medium, end-to-end data transmission will happen with more or less accuracy. With voice recording devices, oral lore passed along from generation to generation may have been just as accurate as written traditions.

End-to-end Communication

Children have an enormous learning capacity. They can be taught sophisticated communication tools through their highly flexible neural assembly. The development of phonemic awareness begins with exposure to phonemic regularities in the form of sounds played repeatedly near a baby's ear[xli]. As a result, the set of neurons dedicated to representing these sounds is activated each time they are heard. Hence, the synaptic connections that connect these neurons strengthen over time, causing the formation of a permanent bond between them. A neural activation pattern is generated when a word is heard, representing a phoneme and nothing else. This process consists of an automatic phoneme recognition structure[xlii].

The formation of awareness begins with the identification and retention of regular patterns in the brain. Phoneme recognition is among the first filters children establish at an early age, allowing them to focus on certain sounds and practically block others from the focus. This process opens the door for language acquisition by establishing word boundaries in spoken speech[xliii].

After the stage of phonemic awareness starts, children progressively develop an ability to pick up and learn morphemes as they grow. Morphemes are crucial to learning a language since they consist of a set of sounds that have specific meanings in the context of that language—semantics. Morphemes are also known as the meaning-bearing units of language[xliv]. In this way, the morphological process in English determines that once a 'single' object is defined, a group of 'several' of the same object is also expressed through the addition of the letter 's'. Thus, the word 'dog' defines 'one' animal with certain characteristic features, and 'dogs' defines several versions of the same mammal.

The acquisition of spoken language adds layers of complexity to the learning process of a human being at an early stage. This intake is manageable for a baby since a healthy brain has sufficient plasticity, memory capacity, and energy to support it. At this point, a child close to turning one-year old begins associating chains of sounds with objects in the physical world. After infants can categorize speech, they learn the auditory forms of words presented in sequences. This stage helps them discover the language's grammar[xlv]. Auditory learning is often the way to acquire semantics and grammatical information gradually. A link between morphological and phonological operations is created at some point in language learning. In this phase, the output of morphological operations is used as inputs to phonological processes to learn words[xlvi]. For instance, for the complex word 'cats' to be meaningfully heard or spoken, the morphemes 'cat' and 's' must be previously known.

Such amalgam leads to speech development as a means of communication between children and the outside world. This critical stage should be addressed: fluency can only be achieved in adulthood if the language is learned before adolescence. After adolescence, the brain begins to reach stability, where resistance to massive neural structural changes appears[xlvii]. Language is one of the essential communication tools for human beings, which places it among the learning priorities for a child. Despite the uncertainties and errors associated with information communication through language, it continues to be the means par excellence for human interaction.

***

Scientific evidence supports the possibility that language affects our perception of the surroundings. Different languages can contribute to shaping brain structures that serve for living in a specific environment. As Edward Sapir found in his investigations, a particular society's language may cause its members to focus on specific aspects of their environment more than others. This hypothesis of linguistic relativity is known as the Sapir-Whorf hypothesis[xlviii].

A study conducted by the MIT found that in Russian culture, the perception (or realization) of the existence of specific colors is translated into the Russian language with great precision[xlix]. An obligatory distinction between lighter blues (*goluboy*) and darker blues (*siniy*) is established—and it has no equal in English. This distinction in their language allows Russian speakers to distinguish between colors that belong to different color categories expeditiously.

The Eskimo-Aleut languages (made up of a family of Alaskan languages) were developed in a northern region of the planet where snow falls approximately seven months a year, consequently leading them to categorize around 50 types of snow with 50 different names. All kinds of snow shapes Eskimo-Aleut speakers see are classified with as much precision as their environment requires[l]. The inquisitive attention to these details allows them to identify and interpret the various snowfalls in the area automatically. Cases like this are found in all languages. Different languages focus on describing certain things with more detail according to what their surroundings require. Abbreviations or specific words that substitute sentences in languages may help speed up communication.

Language must be as precise as possible to correctly convey information between community members. This process filters most of what is not relevant to describe the neighborhood. For instance, the inhabitants of Margarita Island in Venezuela do not need to memorize 50 definitions to represent snow. With only one definition, they can understand that such a phenomenon exists in the northern and southern hemispheres and with less intensity on the highest mountains of the Venezuelan Andes.

The lack of proper communication schemes can endanger us all as a species. Without an effective communication medium such as language, the possibilities would drown us, preventing us from communicating competently. Failing to develop any form of language could change how we understand socialization, culture, education, and logical abstractions. Information exchange is there to enable representations of a diversity of circumstances as accurately as possible. Suppose we cannot effectively describe dangerous objects or places to others or indicate where the food and shelter are located. In that case, task distribution and risk assessments among community members can become limited. Homo Sapiens have been able to survive not only because of the expression of cognition but also due to their rich and diverse forms of interaction.

Via information and communication technologies (internet, mobile phones, computers, social networks, etc.), nowadays, global interconnections allow for further enrichment forms of interaction through learning other languages and exchanging perspectives and information of all kinds. Learning new forms of communication supports knowledge expansion and perspective management, where having more notions and concepts about the world strongly impacts how we engage with it.

***

An interconnected community can become a mess in the absence of filters. A given society must agree on the filters that help to make it functional. Filters can come as laws, rules, dogmas, regulations, or any other agreement allowing for valuable, safe feedback among society members. In communication through language, filters are also vital to guiding conversations between individuals to have a beginning and an end systematically. Long conversations that cast data from random sources without rules and purposes can be unproductive as they may never converge to a practical end. In extreme cases, conversations with individuals with disorganized schizophrenia, dementia, or related disorders and conditions can become disorienting due to the presence of drifts between topics during a conversation that might lead to no return to the main point[li]. In this case, data overflow could jam our average adult brain, considering that we are not made to take into account all possibilities at the same time in a consistent way.

Nevertheless, challenging limitations imposed by filters that our societies impose has become an aspiration for many intellectuals and researchers with 'universalist' mindsets. Personalities like that would combine a list of attributes to excel in many fields of knowledge and get there systematically. By the time this book was written, knowledge had achieved a level of complexity that no human could handle. Still, engines like Google Search, Gemini[lii], or chatGPT can classify, sort, and perform complex reasoning tasks data conveniently. Such a bold aspiration of wanting to know everything has existed in some cultures for centuries when wise men or sages could accumulate enough information to advise various sectors in their communities. Following these premises, Aristoteles taught Alexander the Great the knowledge of the time about philosophy, history, politics, natural sciences, and astronomy. In our times, this endeavor is split among several instructors, and degrees of specialization to grasp the basics. Due to the globalization of knowledge via dictionaries and translations, textbooks, and internet access, we can now realize how challenging it is to have a broad worldview without missing fundamental details. Different fields belonging to hard, natural, and social sciences, humanities, engineering, and arts are turning into complex fields where grasping everything with (even) a high level of expertise is nearly impossible.

Universalistic intellectual aspirations reveal our underlying desire to accurately understand our circumstances in our human world. We inspect our present and past circumstances to plan for the future. When planning our day in the morning, we tend to connect it to expected experiences. We try to anticipate how those experiences will be based on the current state of our surroundings and how they compare to similar past situations. Our predictions are usually inaccurate because we need more in-depth information and an understanding of what surrounds us.

Assertive anticipation is among our most desired ambitions and part of our always-improving intuition system.

A model of a circumstance is an abstract description of the surroundings and situation that a brain creates to discern reality and for forecasting purposes. Many variables need to be specified to describe any situation through a model. The less uncontrolled the environment is, the higher the number of variables and parameters we need when building a model that describes its dynamics. As far as reality is concerned, the number of active variables required to realize such a model might be incommensurable. We say it is incommensurable or challenging to count because reality is hard to abstract as it is not isolated. For instance, quantum physicists attempt to describe the world based on how subatomic particles interact. The variables that will help to explain these interactions are probability distributions, wave functions, mass, potential energy, position, velocity, etc. Quantum physicists can describe simple isolated experiments, but their baseline abstractions are hard to link to the macroscopic scenario. Connecting the double split experiment to the following weather forecast or any other practical anticipation of our environment is challenging. The number of variables between these two phenomena would take much work to assess.

Accurate forecasts are difficult as systems and circumstances are not isolated from the rest of the world. For this reason, we usually make predictions in terms of probabilities instead of dealing with presumed certainties. To illustrate, when deciding to travel from Canada to Patagonia in the middle of July, we should consider different weather conditions—Canada is entering summer while Patagonia is starting winter. Adding the weather conditions to our plans is mathematically expressed as adding new variables to the current model of that future event. In any case, such a given model of what we think the world might be incomplete and inaccurate. In practice, all these variables are impossible to process at once by a single person or machine.

For those who love knowledge and learning, the ambition of knowing about everything, as Johann Wolfgang von Goethe described in his book *Faust*, results in an unfortunate limitation; but maybe there is a way around it using AI. AI processors can be connected to our brains as co-processors to help us analyze our environment more accurately in real-time. Considering a large dataset of information from our current surroundings can enable us to find the best action or thought to be vocalized during a discussion.

The Language Way

Despite all methods of communication between humans, language is the best venue to transmit cultural values and strengthen human interactions. An important part of our exchanges through language is founded on the roots of culture. Given our intricate symbology and grammatical rules, achieving such an advanced information exchange mechanism would have been complicated without many actors throughout generations. The agreement behind the use of languages has accompanied us up to now, and there is no indication that this will change in the future. Nowadays, even if our mother tongue is only used in a remote community with a few members, there are ways in which translations can be done so that we can be up-to-date with the rest of the world. Translations have been present since ancient times. For instance, Egyptian hieroglyphs have acquired meaning thanks to the Rosetta Stone. This stone, which contained inscriptions in three scripts (hieroglyphics, demotic, and ancient Greek), was crucial to breaking the code of hieroglyphs, which is a unique type of writing developed by Egyptians more than 5,000 years ago[liii]. Modern versions of this Stone exist as dictionaries, allowing for one-to-one translations that elucidate meaning in languages with higher accuracy.

The transmission of cultural values not only supports the development of social interactions but also enriches each individual's worldview. Traditionally, storytelling is key to the spread of knowledge generation after generation. Oral storytelling contributed to the foundation of society by preserving ideas, beliefs, moral teachings, and understanding of significant phenomena, technology, and myths that could serve as intellectual foundations for others in the future. The subjects can be as complex as scientific observations, but the narratives used to transmit knowledge expedited the process of understanding and learning since it was also entertaining and engaging.

Language also grants us identities in a swarm of creatures similar to us called communities. The combination of personal traits is represented in our community through the expressions of our words and actions. Without the former, conveying insight into life matters could make our actions look arbitrary. The symbology behind language delivers information to others much more effectively than a concatenation of non-verbal actions that point in the same direction. For these symbols to work as substitutes or complements of actions, the causal representation of physical life must exist in the abstract world of languages. Abstractions are constituted of significant features in the actions they represent. These high-level details are compressed into a few symbols.

In the absence of social interactions, language as we know it might not need to exist. Cases like the life of Ildefonso—a deaf Mexican boy who could not develop language—shed light on conclusions of this kind. Ildefonso's deaf parents raised him in a socially isolated environment. Hence, learning any interaction skills to communicate with others rather than his parents, at an early age, was impossible. Ildefonso could not developed effective social interactions with socially acceptable communication skills. There was not dedicated brain space to any formal coding-decoding protocol to understand others and provide feedback. To try to counteract such an underdevelopment when the opportunity was given, Ildefonso relied on his imitation skills (literally mimicking the actions of other people) as a way to adapt to social circumstances as much as possible[liv].

In this case, imitation seemed to be an interesting adaptation tool for Ildefonso. Imitation is a primitive way of transmitting information from one agent (original) to another (copy). The foundation of imitation is to copy or reproduce the actions of others. Mirror-like repetition of movements and sounds makes it easy for people to learn new skills. For instance, a cartoonist begins his learning by imitating other experienced cartoonists. This plan is followed until the beginner reaches a certain level of mastery, setting the path for them to become an independent cartoonist. Independence is achieved once distinctive techniques appear in the imitation process, allowing them to create unique features that lead to originality and innovation. Originality comes from imitating several other artists, accompanied by skills, individualism, and a pinch of randomness in practice. In this way, the apprentice diverges from the artist's work they imitate. A beginner will, therefore, become self-sufficient when able to preclude imitation as the only means to carry out actions. Originality and uniqueness accentuate our features and identities in our social network. Our unique behavior and actions mark our distinctiveness; without that, what are we?

***

The mere transmission of information between people does not imply effective communication. We can quickly copy someone else's actions through imitation without understanding what they mean. We would merely be following in their footsteps without any intellectual support. Imitation without context does not allow for learning, although it can certainly help to memorize actions that need to be automated. Effective communication may require a different venue that adds meaning to actions. Language as a communication means leads the interaction between morphemes, graphemes, and morphemes[lv]. Without morphemes (meaning-bearing units) in language, there would only be a set of sounds that everybody shallowly repeats. Those sounds would not reveal the existence of any underlying causal organization of sounds that have substance, direction, and context. Only scattered sounds that appear here and there can be recognized without sense.

Although absolute imitation is undesirable, people imitate many things from each other. For example, the basics of spoken and written language are a form of imitation through repetition of the actions of others. Spoken words are imitated so that others can catch them. When a spoken word needs to be expressed aloud, the fine modulation of about 100 muscles[lvi] in the motor systems must be effectively similar in all speakers with comparable body features. Otherwise, generating similar sounds and mouth movements that allow lip reading will take work. Each imitated sound corresponds to a unique object or identifiable symbol—phonemes. The magic of the existence of such symbols is that they can be spatiotemporally transmitted from person to person by emulation.

Nevertheless, such a language acquisition theory must be completed since imitation alone does not guarantee language learning. In the song titled "Prisencolinensinainciusol", created by the Italian singer Adriano Celentano, the words used in the lyrics are made to mimic what English sounds like to a non-English speaking audience—but the lyrics are nonsensical. This song shows how imitation alone does not converge into understanding and appropriate communication. Theories of language learning through imitation have been historically controversial. On the other hand, many researchers agree that imitation as a means of repetition aids learning as a necessary but not a sufficient condition[lvii].

***

We may wonder whether a total absence of language can limit the generation of thoughts. If so, could this be linked to reduced neural activity? This question received a detailed answer from Prof. Barbara Tversky's book titled *Mind in Motion* (2019). Tversky shows that weak or null neural activity does not necessarily occur when language is not used for thinking. Indeed, most mental imagery is not based on words but on images[lviii]. The absence of language does not negatively impact our overall brain activity since we can still experience spatial and visual thinking—leading to neural activity in different brain parts unrelated to language. However, the absence of language could affect how we communicate the outcomes of our internal reasoning to other humans. It also could hinder access to essential sources of knowledge.

Counterintuitively, language can sometimes be an obstacle to thinking and learning. In 2015, researchers at the University of Pennsylvania, led by Danielle Bassett, reported longer learning times in people who cannot shut down communication between different parts of their brains unrelated to the task at hand[lix]. When trying to complete a task, some people put too much effort into finishing it, causing them to overthink. Many people also tend to follow descriptive logical lines using words when trying to solve any problem. This method could interfere with the problem-solving process if it does not need the use of language to tackle the problem. Consequently, the time it takes for these people to understand the problem will be longer than others who do not use language as a means.

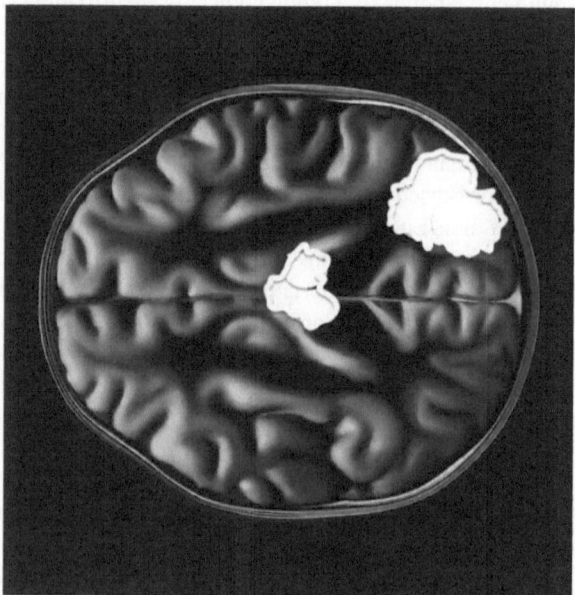

Figure 6: Illustration of localized activity in brain areas when solving a task.

About four years after publishing this study, the authors found that an efficient neural representation of the problem to be solved is indeed linked to effective learning. An effective learner does not require the simultaneous activation of their entire brain to carry out the task: an image-only task does not need the activation of the brain's language regions to be solved. It means that when a person learns to solve a task, well-localized compact neural activities are signs of good performance[lx] (Figure 6).

In any case, language in all its formats (oral, written, braille, etc.) connects us all in our species. It enables stable interactions at a considerably lower speed than quarks using gluon exchange or neurons through electrochemical interconnects. Nevertheless, language may support the emergence of a superorganism called society in which we are simple building blocks. The variety of constituents in various communities may explain the diversity between superorganisms. The evolution of human beings is going slower than the pace at which societies changed from hunter-gatherers to the Industrial Revolution. We might be at the edge of a new stage of humanity with the advancement of AI and the hyper-connectivity of the internet and massive transport systems. Will these conditions of hyper-connectivity at many levels be stable enough to continue to lead to the emergence of something beyond the social organism?

Foundational Flexibility

Simple robustness to external perturbations might not be enough to warrant the survival of living organisms. Other traits are required to keep the stability of a macrosystem. Adaptability brings the flexibility that is needed for the maintenance of life in a given environment. It allows organisms to couple to environments that are native or not to them. Plasticity, as a means of adaptability, is present at many levels in the hierarchical constitution of organisms. To preserve stability and integrity, macro-organisms must slightly readjust their composition in response to a new circumstance. At the molecular level, compounds can temporarily change parts of their structures due to temperature, bind to atoms, or couple to other molecules as the conditions permit. Plasticity allows for reductions or extensions of a body below a catastrophic breaking point. Knowing where that point is can be challenging[lxi]. Beyond the point of fracture, a system loses its identity.

In living organisms, plasticity plays a fundamental role at the genetic level. The modifications that make an entity more suitable to its environment are engraved in molecular compositions defined as DNA sequences. Since genes are primarily composed of DNA molecules, chances are that particularities in their DNA molecular sequence pass on to other generations through any type of reproduction. These changes may be modified by future generations as well. Genetic plasticity happens at an intergenerational level, allowing bodies to enhance their relationship with the environment throughout slow-paced evolution[lxii].

Faster-scale types of plasticity can happen at the phenotypic level, where an organism responds to its environment through changes in its observable attributes[lxiii]. For example, the pattern and coloration of butterfly wings can change as a response to temperature variations or as a camouflage for local predators. This flexibility can make butterflies better at thermoregulation, absorbing or repelling more light, depending on availability. Similarly, light can make plants grow in the direction where the light source is, to absorb it as convenient to them. Not all plants need the same amount of light—species of plants will pursue different venues. Genetic plasticity is like a long-term memory feature, and phenotypic plasticity is short-term memory in terms of information stored in the evolution of species.

An even faster-scale plasticity process happens at the neural level, where the brain self-organizes to represent and make sense of reality as the body browses it. This is how problem-solving develops as the organism interacts with its surroundings. Troubleshooting does not imply finding intelligent strategies to resolve a problem. In many cases, brute-forcing solutions from an ample spectrum of possibilities may also converge to a similar resolution. Trial and error work more efficiently in some cases than in others, depending on how complicated the issue is. Nevertheless, we do not need to implement these trials perpetually. There might be ways to simulate the problem in advance and try to find a solution before going into action.

Nature has many types of dynamics that may pose problems to us. The combination of these dynamics change in time and space over Earth's vast history. Climate change has long term variations due to glacial and interglacial (warm) periods. These changes may have occurred because of the influence of high CO_2 levels in the athmosphere from volcanic activity, vigorous techtonic activity that can affect athmospheric patterns, or the distribution of sunlight on Earth influenced by Milankovitch cycles (Figure 7), where orbit parameters can change, leading to ice or warm ages[lxiv].

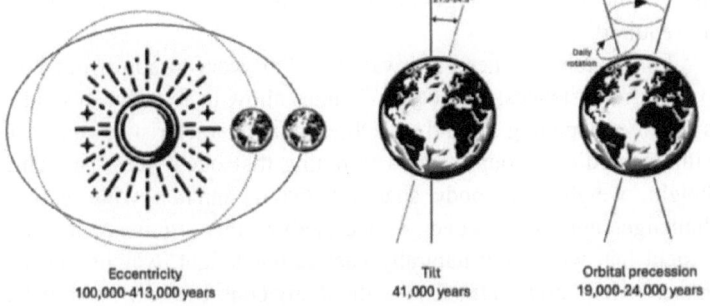

| Eccentricity | Tilt | Orbital precession |
| 100,000-413,000 years | 41,000 years | 19,000-24,000 years |

Figure 7: Milankovitch cycles (orbital cycles).

Other dynamics on Earth take less time to show up. The four seasons that can be experimented within locations near the poles are also quasi-periodic. While ice ages can happen every 40,000 to 100,000 years, seasons have a one-year cycle each, following the orbits of Earth around the sun. At the same time, our planet keeps orbiting around its axis, leading to a half-day cycle of lighter or darker conditions. At the level of weather dynamics, many more factors play roles in their changes. The topography and geography of the location, atmospheric pressure, greenhouse gases, ocean currents, Earth tilt and rotation are some factors correlated to the type of weather within a season.

If we plot the time scales of each phenomenon described above, we will see shapes of waves over time that can be periodic or not. Seasons are so far cyclic, but local weather overcast will vary significantly even within the same season over years of measurement. Adaptations of living individuals will definitely be different for every scenario. Adaptability skills will be substantially dependent on the time scale at which the dynamics occur. A body handling such adjustments needs to be capable of responding efficiently in limited spatiotemporal conditions.

A body needs to be prepared to interact with the high-frequency varying features of the environment. Weather, land conditions, and interactions with other living creatures vary rapidly compared to seasons or ice ages. Slower-frequency variations of the environment can be dealt with through genomics. Genotypic changes can happen as slowly as the eras of Earth pass by. Currently, anthropologists believe that climate change could have transformed the forest of southern Africa into a savannah, which ultimately contributed to the appearance of Homo Sapiens on Earth more than 200,000 years ago. For a new species to appear, fundamental changes in the gene instructions present in the DNA are required.

Fortunately, we need not wait 100,000 years to adjust our bodies to our circumstances. Phenotype changes allow us to adapt to whatever is locally happening around us. Observable characteristics can change without modifying our DNA. The tuning that our bodies can achieve (height, weight, and body shape) helps us interact with reality as challenges appear. However, as we analyze this argument, it may be evident that we cannot naturally change our height, weight, and body shape from an hour to the next without any consequence. Usually, these things take time—even after surgery. At this point, we may ask, what are the attributes that we possess that can be tuned in real-time? Definitely, we cannot tune our entire body composition. Instead, we may have some other hardware that handles real-time dynamics.

Some neural circuits change as the situation requires, representing information in the brain and responding to it automatically or after some processing. Significant microstructural changes at the synaptic level can stabilize after a few hours of learning or memorization[lxv]. This highly tunable landscape deals with some of the high-frequency dynamics that life imposes. But even perfecting these capabilities can take some time from the day we are born. Unlike our tactile sensitivity or motion detection, interactions with reality are not all automated. How we interact with things can require sophisticated information processing, just like communication through language or abstract thinking in images. Short- or long-term memorization also requires critical tuning of neural networks in the brain.

The exploration of our surroundings is bidirectional. Our responses to the challenges the environment poses to us will influence it. But only one system needs more tuning and adaptability than the other one. Humans need to adapt to the environment, but as some of its changes are unpredictable, we tend to modify it as convenient. Instead of predicting exactly when a catastrophe will happen, we try to prevent it using artificial infrastructures. For example, instead of waiting for the rain to stop to do outdoor activities, we designed shelters that protect us from the rain from the start. Instead of waiting for the right season to sow and reap, we have developed food technology and greenhouses where food grows even in winter.

Our actions are dictated by the way we process information. As the complexity of the data gets incrementally high and we are time-constrained to act upon it, this processing needs to be done efficiently. In many cases, we do not have the luxury of using brute force to find the optimal solution to the issues we see on our way. The need for optimal decision-making and forecasting has made us develop techniques that helped perfect these fields.

Decision Making

Decision-making is a process that requires the anticipation of the future as well as the evaluation of past events related to the decision that needs to be made. In adulthood, this capability is mastered once the prefrontal cortex reaches a point of great maturity. At this point, the social behavior of human beings becomes more stable and less risk-taker[lxvi].

According to the literature on decision-making, this process is composed of three categories: (i) decision goals, (ii) a set of various choices, and (iii) a set of strategic instructions or selection criteria. In 2007, researchers Y. Wang and G. Ruhe published a mathematical model to carry out decision-making based on the steps mentioned above[lxvii]. A decision-making algorithm might not be strict because of difficulties in understanding the choices and setting a strategy. To this algorithm, we could also add an intermedium step where "consequences due to possible future actions to be taken" are defined. Explaining the consequences of actions requires an appendix with a collection of choices that are necessary to understand them. Such a collection consists of a list of past events or memorized scenarios that overlap with the decision to be made. Elements common to those memories can be a base for forecasting consequences expected from similar activities. The result of this analysis will lead to weight consequences based on past experiences. The output of these exercises usually converges into successful decision-making[lxviii].

Let's think about the case of a person learning how to play chess. Given that chess is a game of perfect information, with no secrets between players, defining goals, studying strategies, and courses of action based on them can lead to success[lxix]. After several trials, a chess player could find regularities that would allow them to forecast the consequences of future actions in each game. The conjunction of all these experiences will help players evaluate what decision to make when facing a new opponent and the associated risks. Such a reservoir of experiences can only be built over time.

What it takes to make good decisions does not have a unique answer. Getting more experience to reduce risks will always be necessary since uncertainty is continuously present in real-life situations. This is why our brains are not entirely defined and static—they must accommodate modifications. Our neural networks allow for incorporating elements that can reduce uncertainty as much as possible. Despite the existence of such neuroplasticity, there is a point where the brain cannot dramatically change anymore to keep reducing environmental uncertainty in an efficient manner. Some regions of neural circuits lose around 50% of synaptic connections during adolescence, allowing the leftovers to operate more efficiently[lxx]. With the establishment of an efficient neural architecture in adulthood, the brain reaches a state of maturity in which groups of neurons can contribute competently to the performance of stable brain functions[lxxi].

As we improve our decision-making capabilities, the outside world keeps changing, and the regular patterns we have discovered may eventually change. The good news is that our neural architectures are already trained to find patterns in new data by matching new and old patterns stored in memory. As soon as those brains understand how new data relates to past data, their actions will respond to incoming events likely to happen according to estimated future projections. If relations between old and new data cannot be found, the brain might need to learn how to analyze it to understand its composition or collect more associated data to make sense of it. The former case applies when a new intimidating opponent in chess appears. Since this opponent might have a remarkably different game strategy concerning other known players, rewiring the neural networks could be triggered.

In general, the particular shape of each brain is built based on the configuration it must adopt to navigate circumstances. Due to the existence of plasticity, our brains are efficient at analyzing data. Plasticity makes us reliable and general-purpose organic machines—despite inaccuracies.

***

The efficiency gained at data processing is mainly dependent on past experiences. Such information is stored in a cloud, defined as long-term memory. This fascinating structure is embedded in a diversity of brain parts. Although the exact location of memory is unclear, it is possible to trace related brain circuits. This process consists of labeling and tracing memory cells called Engram cells in three parts of the brain: the prefrontal cortex, the hippocampus, and the basolateral amygdala[lxxii]. Once a memory is consolidated, the information stored is ready to be retrieved and used for executing cognitive tasks that depend on it.

As curious individuals, we wonder if a remarkable ability to memorize could be helpful in a decision-making process. A good system could match new patterns we perceive with old ones stored in memory. Such matching would increase the chances of forecasting future consequences, leading us to make the best decision based on what we believe will happen. This possibility appears coherent, but a deeper analysis uncovers a crucial flaw: decision-making is not a single-pole operation. A great memory capacity is desirable if accompanied by an equally exceptional ability to find ways to use what we have stored there. Therefore, a chess player who only memorizes games but does not create strategies to tackle new problems will most likely not have high success in competitions (Figure 8).

Figure 8: A chess board.

Unlike chess, life has more variety in problems to solve and strategies to create. And we are constantly forced to make decisions. Therefore, action plans are always going to be required. Our lives are surrounded by uncertainties everywhere, with an overflow of unknown problems. Using our brains as general-purpose machines has been useful in reducing environmental unpredictabilities. The more we understand our reality—and what to expect from it—the less uncertain it will look to us.

The toolkit for exploring and getting the most out of our surroundings is linked to cognitive development. Cognition has a genetic predisposition that contributes to its development. Genetics influences cognition for more than 50% and up to 70%, depending on the environmental conditions. Socio-economically advantaged environments have been associated with a maximization of genetic influences up to 70%[lxxiii]. These cognitive tendencies can be believed to be talents acquired from birth. Some brains are predetermined to process particular types of information better than others. Our talents are not the result of a completely random process; on the contrary, they are influenced by inheritance. However, we currently cannot determine what talents heirs will be born with. The exact prediction about how a 'predetermined' neural structure will look is currently impossible. Counterintuitively, predetermined structures cannot be determined beforehand, so we can only guess.

Memory

For a decision-making process to be adequate, the source of information supporting it must be reliable. Knowledge and experiences we memorize should be an always-available dataset to facilitate the process.

It can be disappointing to learn that our memory is not as accurate as we think. In fact, our memory has difficulty pinning down the exact details we try to remember—especially if they have happened only once in an emotional scenario. Despite our memory inaccuracy, people can still use it as a source to make valid day-to-day decisions. The lack of accuracy in memories is associated with distortions of data stored long ago. At first, much of the information stored in memory is highly detailed and specific. Then, it becomes more generalized over time. The strategies to retain information in the brain are defined as episodic and semantic memories[lxxiv].

Long-lasting memories are created and consolidated from repeating experiences, actions, or data. Repetitions are essential because experience over time helps intensify neural connections in which learned information is represented. Retaining such information can decline if training is inconsistent or has a prolonged interruption[lxxv]. Depending on the number of experiences, the retained data can be more efficiently and accurately retrieved[lxxvi]. Memory accuracy typically worsens over time, leading to misrepresenting the actual experiences in the brain. For instance, memories of US citizens related to the terrorist attacks on September 11, 2001, were found to change by 37% after one year and 43% in three years[lxxvii].

From small specific details to big chunks of data, so much information could be missing when attempting to retrieve information from memory. Therefore, at the time of evocation, we often create scenes composed partially of elements from memory and partially of similar unrelated information that was mistakenly connected to the original scene[lxxiii].

Reconstructing the past gives the impression that memories are not kept in isolated, independent pieces but are interacting with other elements. We feel that what we remember is right because we believe in its *plausibility*, even though many connectors we use to create the scene do not correspond to the associated memory.

When we try to remember a meaningful conversation, we retrieve the most relevant parts of it to reconstruct the scene. These parts are initially stored as episodic memories and then reinforced through repetition or emotional arousal. Neural circuitries representing such data in the brain gradually strengthen their connections until they are reinforced in memory. At the moment of retrieval, we tend to add details that were not there to get the feeling we correctly remember the conversation, even if it is inaccurate.

As the years go by, memories become more complex. What we remember from a particular event is now immersed in a basin of mixed—new and old—facts. Any given object or experience that is remembered is also being reconstructed. In this step, memorized information is most prone to be inadequately evoked.

***

The notion of "a past reality" in our heads is accompanied by errors generated by the process of reconstruction associated with evocation. These errors add more uncertainty to decision-making processes since they rely on the accuracy of past experiences. When making a crucial decision based on remote memories, we could make a better decision if we adequately recall those experiences. Whatever we recall could be a mere interpretation of what reality was. Interpretation is critical because it can affect the reconstruction process that a person can make at a given time. Along with the interpretation, the reconstruction of a scene could result in a different memory of what happened in the past.

As we reconstruct a given past event, we start incorporating it into our decision-making process to support or decline arguments before concluding. For instance, a discussion between friends could bring up long-term stored memories. During the discussion, some memories have been reconstructed to favor (or disfavor) certain propositions. At some point, one person's more persuasive viewpoint could modify their opponent's evocation, thus generating a type of manipulated collective memory. Distortions can be the result of interactions with misleading information. This phenomenon is called the "misinformation effect". Memory distortions appear when an original memory is exposed to misleading information related to that memory[lxxviii]. Many techniques can be used to feed misinformation to individuals. This way, leading questions could distort a witness's memory during an interrogatory.

Similarly, political leaders could use memories with new reinterpretations that will be useful for their purposes. The collective memory can be reconstructed with corrupted insights that make the new story plausible and sensical. What is evoked becomes part of a new story, which shares only a few elements with the past reality. The seeming plausibility of the story gives a sense of faithful reconstruction of the past. This is where deception comes into play. A coherent and plausible reconstructed story does not have to be true to reality as long as enough people believe in it.

Just as satisfaction plays an essential role in the final shape of an evocation, memories could be influenced by our emotions and bodily states. Emotions are mental and corporal feeling states that contribute to the accuracy of recalls. People under conditions of high arousal (happiness) are found to have substantial memory distortions[lxxix]. Happy moods favor the generation of addition of elements—related to feelings—that do not belong to the available stored data. Misleading information is more likely to be assimilated by the individual and incorporated as false details into their memory[lxxx]. The association between misleading data and memory could modify past experiences in memory, differing from what they originally were. On the other hand, false recalls seem to be reduced in people with negative moods and low arousal.

Mood can also affect the way we process information because we rely on memories to do so. It can lead us to make evaluative judgments based on heuristics that can be disproportionally misleading. Mood states can vary daily in the same way we see a Rubin vase—one day, it appears to be a vase, and the other day, the shape of two faces. Such contrasts allow for the existence of judgmental errors that can influence our behavior when interacting with other people or reacting to our surroundings.

***

Memories can help us understand our present reality and forecast future consequences. Our capacity to travel through time between memories helps us find similarities between complete past and incomplete present experiences. We can discover underlying associations between current and past events thanks to correlations between memories and our prevailing reality. Such associations allow us to grasp the future consequences of what we experience; this prepares us to face them readily. Without this connection, moving forward as a species would be difficult. Spanish poet and philosopher Jorge Augustin Nicolas Ruiz de Santayana claimed that "whoever forgets his history is condemned to repeat it". Acknowledging our past experiences should prevent us from getting stuck in the same problems in the future.

Our brains seem to be assembling objects often. We not only reconstruct from far in the past, but we also constantly construct incomplete present realities to forecast consequences. Correlations between past and present events can be used to approximate our current reality based on what we have stored in our memories. In this "travel through time", our minds explore complex in-memory experiences. This immersive adventure to the past can be lived and felt like a virtual reality or a simulation because it disconnects us from our physical actuality for a moment.

Often, memory retrieval is influenced by emotional interactions through the Amygdala[lxxxi]. The Amygdala is the brain region that plays a key role in emotional processing. Therefore, it contributes to memorizing emotional events more accurately[lxxxii]. Such a process involves interactions between different brain regions related to recollecting the spatial and temporal context of emotional experiences[lxxxiii]. Affective memories can be evoked vividly if we consider that they are accompanied by representations of sensory cues and the overall context of the memorized event. However, some types of reminiscences, evoked entirely as semantic memories, can be less immersive, but helpful. They could help us quickly remember things we need at a given time. Although non-immersive, these memories can also take part in the construction process of the present reality.

We also have skills and techniques stored in memory. This is why once we have trained to ride a bicycle, our brain can help the rest of the body remember what to do when we need to do it, even after a long time. Procedural memory is the specialized unit for storing this type of information. In the cerebellum, procedural memories are generated for any activity that needs to be automatically performed. Hence, when we try to ride a bike again after a long time, it would take us less time and energy to retrieve from memory the steps to do it again than to learn it from scratch[lxxxiv].

Due to the existence of all those different types of memories, some researchers think that the mind's functions can be interpreted as the software run on our organic hardware—the brain circuitry. Considering this analogy, when we compare our brains with computers, it makes sense since we can browse past files from old folders, install programs, and display movies in abstraction.

Memory Dynamics

Memory is not static. It evolves over time as it must connect the past, present, and future. This evolution not only supports making sense of reality but also helps to forecast the future, which allows for planning accordingly. The complex features of reality made us evolve to have different types of memory. A broad list can be made based only on a diversity of functionalities, but generally speaking, memory can be divided into short-term, working, and long-term memories[lxxxv]. Short-term memory stores data we need to use for approximately a few seconds. It allows us to hold data for a short period until we complete a given task. Both short-term and working memories can be seen as spans of immediate memory. The working memory will keep the information for a few minutes if it is valuable and relevant to the task. This memory operates when reading or remembering a verification code or a phone number before dialing[lxxxvi].

If data initially stored in short-term or working memory gets rehearsed or reused many more times for other tasks, chances are it can get stored in long-term memory. This happens when a large set of neurons fire repeatedly, allowing the nerve connections to be strengthened. Once the synaptic connections have been built up, memory traces get reinforced, and their information gets recorded. At this point, memory has been consolidated through a process that can take hours or days, and data stored there can be retrieved at any time. Recalled memories can vary depending on how they were saved. As mentioned, memory recall can be inaccurate as its reconstruction process involves many different sources. Each source is related to the various parts of the brain where data is. As these elements are stored in a decentralized manner, recalling can become a complex operation involving simultaneous interactions with many parts of the brain. The various components involved in this process may also explain why we improperly evoke many events.

Despite how blurry such memories can possibly be, we can reconsolidate them with new information. This process can help to reconstruct the memory better—or, in some cases, worsen it if the new data is not related to the memory in place. Memories can also be formed from interactions between various sources with common elements. Our brains may confuse and mix information because of their similarity, even if data comes from separate events. Without appropriate filters, our minds can associate those sources with the same event. Let's say a person has two memories of spending time at the beach with friends.

Two incidents that happened at two different times could be remembered as if they happened on the same day due to the similar scenarios in which both memories occurred.

* * *

The process of storing data in memory is unique to each person. Additionally, every individual collects specific data throughout their life and remembers it slightly differently each time due to distortions in stored memories acquired over time or issues during reconstruction. In this way, a given event can be remembered in entirely different ways by oneself or among several people. To match these occurrences, we can use synchronization to find a common reconstruction of what happened. For these dynamics to happen, people must interactively adjust each other's stories by accepting or rejecting parts of them. Sometimes, people get synchronized through such acts as imitating facial expressions, by matching breathing rhythms or other physical gestures[lxxxvii]. Other times, it could be a matter of following the flow of the conversation until an agreement is reached.

In general, remembering is being aware of an event or experience that we already lived in the past—as stated by Endel Tulving in Memory and Consciousness (1985)[lxxxviii]. The attention that we pay to such events can immerse us in a somewhat imaginary state. We sometimes feel that we live in an illusion due to all the errors accumulated by our distorted evocations. The constant improvements we try to make to our memories during reconstruction and the feedback we receive from the exterior keep us grounded. Our memories serve to make correct decisions and forecast future consequences; therefore, their appropriate reconstruction is essential for our survival and well-being.

The Forecaster

Information exchange between fundamental particles, atoms, molecules, neurons, and other cells has a significant role in the building blocks of large multicellular organisms and general life's origin. The emergence of organisms cannot exist without any information trading mechanism, where elements can become, in a way, 'aware' of the existence of the others around them. Data transmission between elements supports updates on their current states. These exchanges' strength, extension, speed, and stability will delimit the constraints required to develop internal dynamics. In neural networks, the power of synaptic interactions, the extension of the overall network, and the speed of information exchange between brain parts are the elements that help define a large portion of cognitive capacity. This capacity is the result of the internal dynamics unfolding of neural configurations[lxxxix].

At a higher level, information exchange between multicellular organisms continues. As entities do not exist in isolation, even large living creatures have centralized communication mechanisms for sharing data. As described in previous chapters, language has made humans successful in data trading. It has enabled cooperation and stimulated competition among parties. The communication of states of mind, needs, emotions, lies, trues, interpretations, definitions, concepts, tergiversations, knowledge, myths, beliefs, technology, science, history, politics, morals, environmental challenges, etc., connects the world of the people involved.

The packages of complex information we disseminate through language do not come from everywhere in our organisms. These elements are represented in the micro-cosmos composed of neural networks in our heads. With enough resources to create representations of reality that we can communicate, internal simulations of the world are run in the existing infrastructure that neural networks compose. Thankfully, we can run trial simulations through thinking before action takes place[xc].

Reasoning and consequent decision-making are found to be part of these routines. By leveraging predictive capabilities along with an arsenal of other cognitive tools, we can ensure the stability and functionality of our organisms by forecasting and mitigating catastrophic events that could dismantle them. However, the hardware where these simulations are run is more than a mere hardwired space where information flows like water in the Venice canals. Learning, memories, and experiences also shape this space by updating the architecture of the net.

Complex experiences are squeezed into simple mental models that will never match the objective reality, i.e., the reality is independent of our perception. The cosmos created by neuron assemblies inside the skull has enough dimensionality and resources to host complex representations of the world. Such representations have the shape of simulations that allow us to construct hypotheses and estimate consequences based on the arsenal of data that we use in the analytic stages. The infrastructure in which reconstructions are run is not random. Neural networks that enable simulations have a structure shaped by the environment and learning, providing context to new representations projected there. Results from such runs support generalizing, inductions, deductions, and extrapolations for forecasting.

One of the most significant challenges we face as individuals is finding accurate ways to generalize and forecast events with high confidence. Counterintuitively, we have already developed forecasting skills, especially when risks are posed to survival. Predicting the future is not material for science fiction only. It is a task we do quite regularly. This is an essential feature in our arsenal because we often want to know what is next in order to organize our lives accordingly. We do our best to estimate the future consequences of actions or forecast the weather, stock market, food availability, and geopolitical scenarios. Predictions are not made randomly, as they depend on analytic or empirical models of the subject. Such models mainly encompass general or specific aspects of our circumstances.

How we build models of the world can vary from person to person. Since personal models are usually not rigorous enough to become universals, people never share them verbatim, even though the environment may be the same. For each of us, the world is different because we perceive it and feel it like that. The context will determine the skills required to build models that support accurate generalizations from any given dataset. In this way, Pre-Socratic philosophers or naturalists theorized about nature. Thales de Miletus, Anaximander, Anaximenes,

Heraclitus, and Empedocles laid the groundwork to understand the nature of matter based on a few generalizations about the fundamental substances that compose it. Each philosopher proposed a natural element or a combination of them (water, Apeiron, air, fire, and earth) to explain the nature of matter[xci].

Models of the world attempt to describe what happens around us and beyond[xcii]. Causal explanations of everything are not limited to what we can observe or measure. Faith in supernatural explanations is at the core of our understanding, especially if no other justification can be found using physical resources. For this reason, faith and religion are as old as human attempts to make sense of whatever patterns can be found in nature. Despite the importance of faith, our primary focus in this discussion will be on elements that adjoin knowledge with what can be tested. Religious experiences and reasoning based on faith are challenging to fit into causal models that can help forecast what will happen objectively.

In general, causal models should be tested or confronted with nature[xciii]. The best way to test the validity of causal models is by using them to predict the future. For example, a model of the weather would better use mathematical tools on data to predict when the next snowstorm will occur in the coming 15 days. The best model depends on the amount and quality of data used in the analysis and the careful utilization of tools (probability and statistics, machine learning, etc.) to process it. Once a model has been developed, it will fit the provided data and determine the possible future behavior of the weather.

Models that we use daily may emerge from learning, imagination, beliefs, emotions, and feelings. Active interactions with the environment make us seek an understanding of the various phenomena coming from it. Models help us understand our present by embedding it into a causal template and creating a prospect of how the future will look based on extrapolations of such causal constructions. Discovering the causes of what happens is a big part of what keeps us as stable entities. Not being able to relate causes to effects might endanger our survival. Indeed, we have brain infrastructures to make associations that allow us to avoid disturbances beyond what the stable balance of our body can deal with. As such, we commit to ideas when planning for the summer holidays while speculating about potential future conditions. As a result, we avoid picking a holiday destination in a location where an armed conflict is predicted to continue in the near future. This is one of the reasons why tourism in Israel decreased considerably after the war with Hamas started in 2023. Opinions, biases, and forecasts are all parts of individual models of the world that can often converge into a common conclusion when shared with peers.

Nevertheless, if not enough information is provided when building these models, they can be deemed to be highly inaccurate. This is expected, as we cannot expect to have clear ideas on subjects almost wholly unknown to us. Without good models, we are challenged to have misleading opinions about topics we do not know. Even though nobody should be encouraged to have a say in matters that are unknown to them, some people are still forced to do so, often creating more problems than not. We all have faced this type of case, haven't we?

Creating causal models of reality that are 100% accurate is so far impossible. Instead, we deal with approximations of it. Whatever we want to forecast about the future of our complex surroundings will always be accompanied by a measure of the likelihood of it happening. Unless what we wish to predict belongs to a well-known controlled environment, its future is conditional to an assortment of hidden variables. Our thirst for data and control to serve our models has made us achieve unheard-of goals that no other mammal species has achieved. Since we are permanently at the disposal of what the planet Earth internally does or what outer space brings, our civilization has sent explorer rockets to space, put satellites in orbit of Earth, and added cameras in a diversity of uncomfortable places—hopefully, as the legal limits permit—to watch those dynamics. Unfortunately, this data is necessary but not sufficient to grant 100% accuracy to our models.

Conditional features like that have taught us that our models should always be tunable and adaptive to whatever changes that may occur in the environment. Consequentially, the success of our causal models depends on their adaptability. As current AI engines demonstrate, adaptation is the way to go, but let's elaborate on this proposition. In the past, the mathematical models developed by scientists for data classification, pattern recognition, and clustering could not adapt to the ample complexity of the data. Scientists have always known that humans classify, recognize, and cluster data better than Turing-Von Neumann computers[xciv]. Humans need those well-tuned skills for field analysis, decision-making, conveying ideas, creating art, and anticipation. Therefore, to successfully complete those tasks, artificial engines with human-like capabilities should come to exist. At present, the field of AI is making impressive progress in part thanks to the adaptability of the new neuromorphic models used for it. Models like GPT, GANs, attention networks, etc.[xcv] can be tuned using real-world data to solve problems like text generation, face, voice, and handwriting recognition, videogame playing, art creation, the anticipation of the future, and many other human-like tasks. Adaptability has proven to be behind the probabilistic nature that makes our causal models less robust and deterministic, but more successful in practice.

Interestingly, probability is not the best skill that we have tuned. Scientists have found that people struggle to solve problems with a probabilistic foundation. That is the case of the paradox of Monty Hall[xcvi], which is a mathematical problem that defies the intuition of even experts in the area. In 1990, Marilyn vos Savant shared a puzzle in *Parade* magazine with three closed doors with goats behind some of them. The contestant must select one door to see if a goat is behind (Figure 9). Right before that door is opened, the host opens another door revealing a goat inside and then asks the contestants if they should stick to their original choice or choose the other unopened door. Savant's solution is to switch to the other unopened door because the probability that the goat is behind doubles. This choice became a scandal among mathematicians who estimated that sticking to the original choice was a better strategy. Ten thousand letters and around a thousand signatures of protest were sent to vos Savant from prestigious academics that were eventually proven wrong. This example shows how our brain fails at solving probabilistic puzzles, which is hard even for specialists.

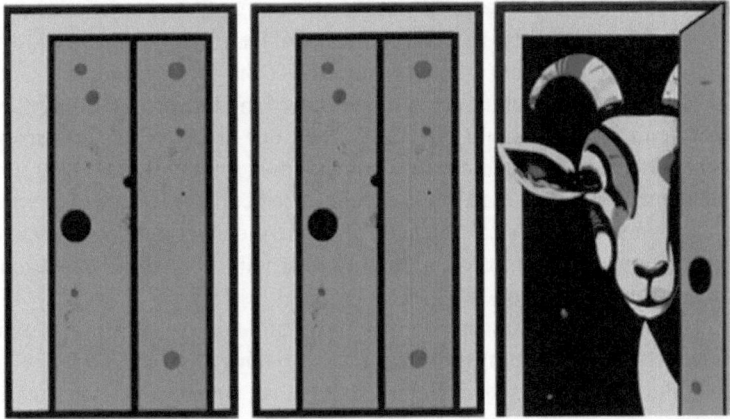

Figure 9: Monty Hall problem.

Models that work are the most adaptive, and based on probabilities rather than certainties. With high adaptability, if models are wrong, they can be modified with less resistance than a hardwired structure. One of the struggles posed by this flexibility will always be the thirst for patterns. Finding patterns in data is almost instinctive to us, but we do not always do it right. We tend to generalize from those regularities no matter how small or large the dataset is. There might be a primitive need behind it. The limited time we have to make decisions based on the patterns we can recognize and then act upon confrontation pushes us to consider binary (yes or not, true or false) options only. This patterning instinct has made us become believers in models or ideas that may not match the causal developmental line of reality. However, those patterns may have enough resources from the environment to convince us of · their accuracy. Ideologies, religions, faith, hunches, intuitions, omens, prognostication, and superstitions come from common patterning needs. Understanding the needs behind these concepts encourages us to develop better models to navigate our environments more carefully.

***

A causal model seeks to find reasons for things happening. Causes are the reasons why effects show up. If we can identify the reasons behind every event in the universe, we can determine precisely what will happen next. This thought experiment is known as the Laplace's demon, which the French scientist Pierre Laplace proposed[xcvii]. Deterministic laws of classical physics lead to the total prediction of the future if they are known. We can also retrodict (the past) and predict the future using the same laws. The laws of nature are the skeleton of the universe, the interface that connects causes and effects. Beyond our objectives as a species, our current understanding of classical (macroscopic) and quantum phenomena does not lead to the accurate prediction of what will happen in the universe. Physical sciences offer theories and methodologies to describe what we observe and predict what will occur if the conditions are known and controlled.

Nevertheless, humans do not have the luxury of waiting until scientists find deterministic laws for everything that add certainty to the future—if they existed. Beyond faith in the supernatural, the truth is that we are on our own, and we must make decisions as we go. Instead of generating causally deterministic models of the world, we accept the probabilistic nature of what those models can achieve, conditioned to hidden variables. In practice, we navigate life by considering the rough probability of something happening. Probabilities are all we get. For instance, we can estimate the likelihood that hikers find predatory black bears on a random trail in northern Ontario (Canada). After going through the data, we may conclude that the probability is low since these are considered "extremely rare" events—only approximately 1.5 black bear attacks have been reported in North America since 2000[xcviii].

In more urban settings, challenges require us to be more prepared for many more probabilistic future scenarios. Since the moment a person leaves their home to go to work in a cosmopolitan city like Montreal, it is tough to predict what can happen. On the way, the subway may or may not work, have delays, be jam-packed for hours at some popular stations, and have unreliable schedules. Due to related challenges, morning commuters can become moody, reactive, and pessimistic. Once done, the commuter is prone to feeling overwhelmed already, but the day has just started, and there is a need to deal with the workplace context next. At work, this person may need to solve engineering problems, repair a computer, assist patients, deal with a terrible family law case, or any other activity that requires high focus, problem-solving resources, and creativity. Cognitive and emotional capabilities should be well-tuned to succeed in this calamity. We may wonder here what makes us deal with our realities better than others. Are there specific traits that support more efficient navigation of life and better utilization of resources available to do so? Can those features be objectively measured and enhanced? Let us examine how cognitive capabilities play roles in the development of models of the world.

Intelligence

To have complex models of the world in our minds, we need the proper infrastructure to represent them. Nothing surprising up to this point. It turns out that, compared to other animals or members of the hominid family, Homo Sapiens have a sizeable cerebral cortex that is populated by a generous number of neurons. We know that because, nowadays, magnetoencephalograms allow scientists to obtain neuroimages in the millisecond time scale[xcix]. This means that the information flow in the brain can be followed with higher precision. One of the most significant discoveries this technique has uncovered is the role of white matter in information processing. It emerged that a white matter bundle named "arcuate fasciculus" is connecting regions in different brain parts[c]. Such white fibers can be interpreted as a highway where data travels at a certain speed. As data inputs the brain through areas connected to the sensory system, it travels forward, passing across several regions that perform some processing. At each stage, information is processed, which may join a decision-making routine and then stimulate motor/speech areas.

The speed at which information is transmitted through such a highway can be related to efficiency in cognitive tasks like real-time problem-solving or future anticipation. Although neuroscientists have not yet proven this, much has been speculated about the cause associated with different intelligence levels among people. In any case, processing speed is fundamental to support the development of skills required to solve a complex problem. Suppose the problem involves the interaction of skills like verbal comprehension, spatiotemporal reasoning, excellent working memory, short reaction time, and decision-making. In that case, paths interconnecting many brain regions are likely required to perform at high speed.

Intelligence encompasses an assortment of measurable skills, such as logical-mathematical, spatial, verbal, reasoning, creative, kinetic, and memory proficiencies[ci]. As something as broad as intelligence is complex to define accurately, researchers in this field choose to include more or less skills to the list that covers intelligence. We are equipped with cognitive instruments that help us understand a problem, troubleshoot, and perform forecasts following some guidance. The concept of intelligence is fundamentally related to the skill set that a specific culture interprets as valuable and that the natural environment considers essential for survival and well-being. Each culture weighs the skills required to achieve objectives in a geographical location[cii]. Culture delimits the lifestyles of social groups that share a constrained geographical region. An established culture comprises values, customs, beliefs, knowledge, and artifacts from constrained collective behavior.

As our biology is shaped by nature, culture, and artificial artifacts emerging from the human collective, we process information influenced by the templates hammered by our circumstances. Context is vital in understanding how the concept of intelligence is structured and valued by communities. From this framework, we can realize how challenging defining intelligence is, so we do not define it, we put together statistically-based concepts instead. Despite the disagreements, what is universal to all that conglomerate is the features set that the psychologist Robert J. Sternberg has identified as the development of strategies to solve problems, monitor the implementation of these strategies, and evaluate their effectiveness. Many of those skills can be enlisted in a grosso modo and measured to determine fingerprints related to individual capabilities[ciii].

We must be aware of a potential excess of simplification on something as complex as human intelligence, as we are deemed to lose information about it. Researchers like Scott B. Kauffman are promoting appreciation into a more holistic, and less reductionistic, measure of intelligence. When we reduce people to single metrics, identity is compromised. Let us delve into the specifications of how intelligence is measured.

IQ

Psychometrics was initially a field focused on testing children with learning issues at school. By identifying these challenges, teachers could uncover the root of the problem. Alfred Binet and Theodore Simon devised a test that measured performance on various cognitive tasks to help children based on the results. Although effective back then for the purpose for which it was created, measuring general intelligence with the Binet-Simon test was not possible for children and adults alike. From this point, psychometrics has evolved to assess what is nowadays called intelligence quotient (IQ), which includes a battery of tests to measure performance in various mental skills. Modern IQ tests are not absolute measurements of intelligence. A person's IQ score only makes sense in the context where it was measured since it is relative to their peers (same culture, age, and gender). These scores are not arbitrary. High scores are correlated to successful lives in the long term and can be used to predict some degree of accomplishment that is culturally fulfilling.

IQ scores are shaped by the bell curve distribution, where the mean is set at 100. If we visualize the shape of a bell, we can see that it has a top and two bottom segments surrounding the top (Figure 10). Most people are included in the region that contains the 100 mark. The bottom areas around <80 and >120 are so far from the average that they are defined as exceptionalities. Exceptionalities are understood as very low (<80) or high (>120) intelligence with respect to their peers. High scores are correlated to more learning efficiency and adaptability. Based on a battery of tests that usually include reasoning, spatial ability, vocabulary, and processing speed, psychometrics intends to cover all possible skills related to life success and even predict lifespan. People with higher IQs are linked to placing themselves in environments and behaviors that promote better health[civ].

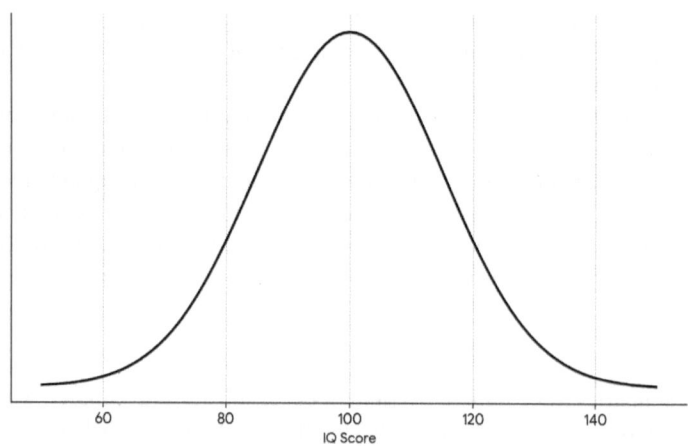

Figure 10: The bell curve distribution.

Where does that IQ score come from? Is low, average, or high intelligence the product of pure genetics, or is the environment responsible for our fate? Both have their share in this matter. Studies on twins' brains using neuroimaging techniques show that genes and the environment contributed to the formation of white matter fibers, i.e., the highways communicating brain parts[cv]. Efficient problem-solving abilities in IQ tests can be linked to better interconnects between brain parts. How the various environmental triggers turn genes on and off for different purposes has yet to be explained by science. We can only wait for more data to find causation out of correlations. What we know for the moment is that we are not determined by our genes as long as the *indeterministic* environment participates actively in our development as individuals.

Good test-taking skills are an asset when evaluating IQ. Naturally, a person with low test-taking anxiety will perform better than anxious takers[cvi]. As discussed in previous chapters, emotions can interfere with memory, and reasoning may be distorted. A way to discard emotions from the scenario is by taking the test multiple times until the IQ score becomes approximately constant. But it may happen that anxiety never goes down. Motivation is also a factor that interferes with maximizing general intelligence. Low motivation for strengthening cognitive skills through study or hard work may negatively reduce the chances of succeeding in life despite the IQ score.

Intellectual abilities can vary between cultures and people with different emotional and bodily states. People with poorly nourished bodies can show cognitive deficiencies. Emotions interfere with our focus and can weaken or strengthen our performances. Some psychologists, like Robert Sternberg, include other features rather than reasoning to encompass all that can intervene in developing a successful life. His triarchic model of intelligence includes analytical reasoning, creativity, and practical skills such as common sense and emotional intelligence[cvii]. Although extremely tough to measure, emotional intelligence should be part of a comprehensive overview of someone's capabilities to succeed in life. People with poor emotional intelligence will not be able to team up appropriately at work or have the correct self-control to crew a rocket and solve problems that may arise along the way.

We must remember that human brains are not literally "reasoning machines". We deal with emotions as well as logic. A long chain of analytical blocks does not solely determine our decision-making but is also a matter of gut reactions. We are still instinctive despite our 1,000 cm^2 of unfolded cortex[cviii].

Uncomfortable with this, logic enthusiasts are working hard to build robots and computers that do not have to deal with emotions when they process data. Granted that even machines may fail to perform as expected due to structural problems, they cannot express emotions regarding these defects. Once fixed, machines will return to perform as they were before the breakdown. Importantly, emotions cannot interfere with their performance. In contrast, humans will continue to think, feel, and live even if reasoning is obtuse. People must thrive no matter what because they must continue living in an uncertain world that requires them to make daily decisions with little information, and it's sometimes hard to shut down and let it be.

Various Commanders

Similar to emotions contributing to our cognitive processes, for better or worse, so do our intestines. The circuitry of sensory neurons in the gut has been speculated to be the "second brain" of our bodies[cix]. These neurons, located in the digestive tract, have their functions controlled by the autonomic nervous system and by hormones. Neural networks that form part of this enteric nervous system can learn and memorize[cx], but they cannot be modified at will to solve math problems as far as we can tell.

The interactions between the enteric nervous system and the central brain happen in a feedback loop, such that our thoughts affect our intestines, and the opposite also happens. An irritable bowel can negatively affect our mood and cause us anxiety, for example. The constant disturbances of the intestines that send signals to the central nervous system can distort how we perceive our circumstances. These disturbances are extra sensations added to our sensory perception and modify our data processing capabilities.

How well or not these distortions can be controlled may impact our cognitive performance and decision-making. The microbial crowd is diverse in the multitude of species inhabiting the human gut. There is a reciprocal benefit to having a gut microbiome. Often, bacteria populate our gut to support the digestion of fibers and some carbohydrates that the stomach enzymes cannot handle. In return, these bacteria take advantage of the leftover fatty acids produced. In addition to helping digestion— which makes us feel better—gut bacteria can produce neurochemicals associated with mood regulation and cognition. Serotonin, dopamine, and melatonin are the excitatory and inhibitory neurotransmitters that reach the brain through the microbiota–gut–brain axis. When these neurotransmitters are imbalanced, a handful of psychological and cognitive disorders may show up. Such an imbalance can lead to anxiety, depression, and learning and memory deficiencies. Therefore, a proper diet rich in neurotransmitters generators can help improve health and cognitive performance. Dietary sources of histamine can be fish, soybean food products, cured dry meat, or dairy products; sources of dopamine are bananas, tomato, spinach, or apples; and sources of serotonin can be coffee, potato, or avocado[cxi].

As the brain is not isolated from the rest of the body, cognitive performance can be affected by all these variables. Nutrition, hormones, feelings, emotions, brain oxygenation, hydration, physical activity, immune responses, and gut health can significantly balance our mental or intellectual performance. If general intelligence was the only marker to characterize our cognitive capabilities, our performances could be comparable to those of robots and AI. But there is a reason why these creatures have yet to reach human capabilities, and that is because we rely on our internal ecosystem to process real-time information as efficiently as we do. But it can also perturb our logical reasoning. Whatever happens in the body influences our thoughts and perspectives—good or bad. Ignoring our internal situation adds random noise to our information processing system. How we handle that latent and impertinent noise that distorts our perception of reality is a challenge everyone must deal with daily.

There is no standard method for dealing with intestinal, hormonal, and emotional noise. Throughout our lives, we learn how to reduce this noise, but on many occasions, it can still distort our reasoning. For example, when we are suffering from food poisoning, our body decompensates, leading to changes in mood, behavior, and even loss of consciousness in the worst cases. The lack of clarity added to our thoughts and perceptions via these perturbations leads us to step back when vital decisions must be made. Such ambiguity makes us evaluate and connect with others to check if our understanding is significantly distorted.

Similarly, when we are upset after an irritating argument with someone else, we find that our way of thinking becomes profoundly distorted. We see that we begin to construct fallacious arguments of all kinds, leading to false conclusions that we are not aware of at that moment. Erroneous conclusions are the first step to making wrong decisions that can be avoided simply by waiting and rethinking them later[cxii].

In general, it is difficult to find the perfect situation where we can make the 'right' decisions without considering our bodily state. Our bodies constantly interact with the outside world; they sense and process various stimuli. Our bodies are never the same at any time in life simply because the external circumstances are not and also because the dynamics of our bodies change over time. Remembering that we depend on various bacteria species to fundamentally process information correctly is essential, so internal dynamics are highly influential in us. Based on these facts, we see that it is almost impossible to analyze circumstances without the intervention of multiple distortions coming from the body. Subjectivity will always be wrapping our conclusions.

The search for objective ways to analyze information leads us to the birth of logic, which we use to determine what is valid and what is fallacious. Although counterintuitive, humans use fallacies during conversations and even when thinking alone. One of the most common misconceptions or faulty reasoning is the hasty generalization fallacy, where claims are made based on a small set of samples. For instance, a person who lives in the Franche-Comte region of France may think that all cows in the country have white and brown colors if they have never visited another part of France or do not have access to the internet or encyclopedias. But as soon as this person discovers the existence of Vichy cows, they realize that cows can also be brown only. This experience will change their first generalization or inference from samples[cxiii], leading them to conclude that there might be more variety in cow colors than they can see.

Reasoning depends on our inductive, abductive, or deductive skills and the information we possess to support that analysis. Fallacies are often built up due to a lack of information, poor reasoning skills, or improper use of the logic laws. In some cases, emotions, the gut, or hormonally unbalanced states can distort our logical thinking and make us generalize too early to get quick certainties.

Dimensions and Superorganisms

A hierarchy of elements in human bodies, from elementary particles to psychosomatic dynamics, has a great diversity. To keep a body in an identifiable piece, some stability in every composition layer must exist. For instance, must atoms composing our organism should not spontaneously split out of the blue. At the particle level, spontaneously splitting an atom into its internal constituents is an arduous task to achieve. To get there, a spontaneous nuclear fission process must occur. Spontaneous fission is not a typical phenomenon in the atoms composing our bodies, as we barely have any very heavy elements involved in our biological processes. Our bodies are composed of lighter elements such as carbon, hydrogen, oxygen, calcium, potassium, sodium, etc. Heavy elements indicate that the atoms have more mass in the nucleus, so a nucleus with a high number of protons and neutrons is heavy because protons and neutrons have mass. Heavy elements, like some uranium or plutonium isotopes, can promptly split their nuclei into smaller, less heavy nuclei[cxiv]. A spontaneous process like that may not be safe for keeping such a large hierarchy, like the human body, in one piece for so long. The effects of an atomic bomb or exposition to ionizing radiation can alter the constitution of the state of atoms in human bodies. This radiation makes the atoms lose or gain electrons around the nuclei, usually leading to biochemical damage.

When atoms form molecules through the interactions between their electrons, certain stability is achieved. At a low level, sharing electron pairs between atoms of the body contributes to the molecules' stability. As mentioned earlier in the book, carbon (a fundamental building block of life) forms strong bonds with elementary molecules (like oxygen, nitrogen, or hydrogen) and carbon atoms, contributing to the robustness of biomolecules' constitutions. At a high level, highly essential molecules like DNA are composed of carbon, oxygen, nitrogen, phosphorous, and hydrogen, which exploit the robustness of the strong bonds created between carbon and other molecules crucial to life. DNA is one of our bodies' most stable molecule complexes since it should preserve genetic information throughout a lifetime. Collagen, keratin, and bone minerals like hydroxyapatite are some examples of other molecular compounds that support the maintenance of structures in the bodies with long-term, robust integrity[cxv].

Since molecules are larger structures than atoms, a wider variety of splitting mechanisms can break down their constitution into smaller units. However, the breakdown of some molecules present in a human body may not lead to any collapse of their integrity. On the contrary, it may be part of its metabolic processes to facilitate digestion or the production of energy in the organism. For instance, digestive enzymes can break down complex molecules from the food we consume so that our bodies can absorb them[cxvi]. Harmful mechanisms such as the exposition to high temperatures can cause thermal decomposition, which can contribute to breaking the atomic bonds of molecules. Radiolysis can also make molecules split through the exposition of high radiation levels[cxvii]. Avoiding exposition to those scenarios may keep us safe at that hierarchy level.

At the level where a cell acquires the property of an identifiable entity out of interactions between molecules, achieving integrity has more paraphernalia. DNA repair mechanisms and consequent quality control are some crucial processes behind forming and preserving cell identities[cxviii]. Errors in DNA replication need to be repaired to guarantee the correct functioning of the structure. If this repair of damaged DNA is not done correctly, mutations (or changes in the DNA sequences) can accumulate, leading to cellular dysfunction and potential cell death. Multicellular organisms can also have catastrophic reactions to uncontrolled mutation, which can be linked to the reproduction of many kinds of cells that are foreign agents to the structure in place. The uncontrolled proliferation of foreign cell batches can contribute to cancer development[cxix].

If we move one step higher in the body hierarchy when groups of cells network together to constitute an organ as fundamental as the heart, a particular dynamic rules. The synchronous movement of cardiac cells keeps the heart working normally[cxx]. From a broader perspective, disruptions in the synchronized contraction of these cells can cause the emergence of arrhythmia, which is associated with an inefficient pump of blood to the rest of the body. Depending on how severe the arrhythmia is, the symptoms felt by the host can go from palpitations to sudden loss of heart function[cxxi]. The body can trigger compensatory mechanisms linked to the autonomic nervous system to restore synchronization[cxxii]. This system may be activated when it detects a decrease in the blood volume that the heart pumps to the rest of the body. Again, depending on the severity of the case, spontaneous compensatory mechanisms may not be enough to re-establish cardiac cell synchronization.

As we climb the ladder of complexity in large multicellular organisms, we realize that the mechanisms that keep internal subsystems together are more sophisticated. We can clearly see that many things must go well for our bodies to work with certain normality. It is impressive how all these mechanisms evolved to coordinate in such a marvelous way in a lifetime. For instance, the relation between neurotransmitter production and cognitive performance shows the coordination of organs in the body. All this happens while the heart pumps blood to these systems, and the respiratory system gets oxygen into the lungs and carbon dioxide out of the lungs. The coordination is only possible by a finely tuned exchange of information between all these organs in the shape of blood, oxygen, glucose, neurotransmitters, synapses, etc.

Apart from homeostasis that supports the coordination of various regulatory procedures in charge of ensuring the proper functioning of our bodies, some higher-order functions contribute to this task. Cognitive capabilities emerged to coordinate external and internal bodily operations altogether. Without a coordinator who can plan where to find sources of energy to provide to the body, find shelters when the temperatures drop considerably or safeguard against predators, it is challenging that such a visibly massive organism could have survived for a long time. To name a few variables involved, providing a continuous food source and safeguarding the body are complex tasks. It takes abstraction of the situation and consequent planning. Such a level of abstraction evolved to the point where the representation of information in the brain became possible. These representations keep fundamental features in the original data that facilitate information processing. Features can be interpreted as patterns that appear in data, containing essential details useful for eventual anticipatory analysis, problem-solving, and decision-making.

Patterning

Humans are pattern seekers in a disordered media[cxxiii]. Patterns provide important clues valuable for the survival and well-being of our species. Regularities in the data we analyze help us adapt and modify the environment around us. Those modifications are often the next natural step when an understanding of the specifications of a given environment has been achieved. Such patterns constitute a base of our empirical knowledge and serve as foundations for making sense of upcoming data. It is expected to find scattered patterns that do not seem to be related. We must make sense of all these invariants to understand what this world is about. The diverse and hazardous things in life and the uncertainty of what surrounds us cause a feeling of disconsolation when we face the possibility of never being truly sure of what is happening.

Patterns do not arise to be perceived. We are the ones who interpret specific dynamics of internal and external structures as patterns. Everything indicates that we evolved to uncover some patterns and skip others. Symmetry is a perceptual feature we appreciate and seek. We also uncover patterns in data that repeat in different scenarios. Patterns that arise in repeating events are significant as they allow us to find correlations and sometimes causation relationships in data. Such causal forms are often constructed from dependencies or correlations we detect in data.

As part of the apparent need to find patterns in data, we tend to extrapolate on those patterns and build cause-effect relationships as soon as possible. To illustrate this point, let us analyze the following expression: "Having a high IQ puts you more at risk of mental illness." The most straightforward pattern we can find in the provided information is that the risk of having mental illness depends on a high IQ but not the other way around. Therefore, being at risk of mental illness does not necessarily imply that a person has a high IQ. More in-depth, we find that having a high IQ does increase a person's chances of mental illness. However, we cannot assume that all people with high IQs will eventually develop mental illnesses. This case is driven by chance, even if the risk is higher than for other population members. Moreover, this statement does not show that high IQ is the cause of mental illnesses; it instead establishes a correlation between both phenomena. Correlational data can become part of a causal explanation when the reasons behind these two phenomena' interactions are uncovered. Correlations are necessary but not sufficient to build up causal explanations. However, correlations allow us to find functional patterns that we incorporate into our daily decision-making and problem-solving.

Uncovering causeless correlations through tentative theories is an important first step. Nevertheless, our theories will never be complete if they are based on statistical analysis only and cannot be explained using deductive reasoning. When a conclusion logically stems from its premises, deductive inference is ruling. Even though deductive reasoning is often impossible to realize because of the lack of foundational knowledge, correlations represented as patterns help us understand the world well enough to sketch models of it and take action.

The case for the establishment of laws of physics is fascinating. Laws of nature mainly arise from sources in which no counterexamples are apparent. Events formalized as laws of nature through induction by scientists are established as their repetition has been consistent since the beginning of time, so there is no reason to think it could be or will be otherwise. From the laws of nature, we build our theories, interpret events, and find related patterns. Rudolf Carnap describes this approach in *Philosophical Foundations of Physics* (1966)[xxiv].

We can describe our adaptation to the environment by identifying specific types of order, or patterns, from understanding the progression of events over time. Nothing exists in isolation. As the world modifies our biology, we modify it back one way or another. We have progressively started changing the natural order of things and adapting them to our requirements. This is an activity that other animals and insects engage with as well. Ants create their own terrariums and hierarchical social structures to stock up on food and supply the queen, who is the generator of the eggs and is responsible for new births. A similar situation happens with bees. Beehives are structures meant to protect the queen bees and its eggs while sustaining a physical structure where the hierarchical society functions.

One of the most significant differences between how humans organize communities and other forms of organization is that we can modify large territories through established hierarchical societies. We can acquire a shared purpose between different human groups and provide the material conditions for this to happen[cxxv]. That common purpose can be led by simple tasks such as hunting together or highly complex tasks like building entire cities or countries. The common purposes guide our building power to appear far more diverse and sophisticated than other species. Some common motives are so extravagant that they lead us to change our environments to carry them out completely. In our modern world, we see how humans have invaded the space of other species at their convenience. Starting from eastern Africa more than 200,000 years ago, the colonization of the planet has been spreading ever since.

As we colonize, some species become extinct, and environments are destroyed. The world and other species cannot adapt fast enough to the accelerated progress carried out by hyperorganisms called communities. At the same time, we are gradually realizing that when we exterminate these species or destroy broad natural spaces, we break the natural balance of our planet. This fact has dire consequences, and we are beginning to be directly and indirectly affected, with the possibility of extinction due to situations we trigger. Currently, many hyperorganisms are trying to reverse what others are doing by helping endangered species survive through many conservation programs around the world. Many institutions provide endangered species with a safe space similar to their natural habitat, in which they can develop under the ever-vigilant gaze of humans. Pro-animal associations and environmentalists need more resources to slow down global extinction. Unfortunately, main hyperorganisms, defined as governments, do not contribute enough as they have other priorities than survival.

Superorganisms

In our interpretation of what is happening with the planet, environments and species alike are having difficulty adapting to the fast pace of the hyperorganisms' dynamics. These hyperorganisms or 'superorganisms' are, composed of interconnected humans, with a shared understanding of reality that might differ from what every individual member thinks about the same subject. This is how we see environmentalists traveling long distances on airplanes to Dubai from everywhere to attend the United Nations Climate Change Conference (COP28) in November 2023[cxxvi]— to name one. Motivated by the gathering, the conference attendees will go there, no matter what, despite knowing that they are individually contributing to the problem by commuting there in the first place. Motivations of the superorganisms may differ from what individuals may do when the organism splits.

Natural resources can be used, but they can also be exhausted. This is one of the reasons that can lead us to pursue an understanding of how nature works to replicate its internal mechanisms that supply our needs. When we build artificial suppliers, we set up controlled environments based on successful rounds of trial and error. In the absence of proper development of causal models of the world, superorganisms may generalize in their analysis. In this way, with the shared purpose of curing diseases, healers in indigenous societies were prone to find medicinal benefits in plants after several rounds of trial. Their methods consisted of finding a persistent regularity in using certain plants. However, these healers did not know the origin of plants' benefits or how those interacted with the diversity of body parts. They did know the plants worked for a specific application.

Correlations are sources of clues to find interactions between mechanisms that govern us, but sometimes they can be troubling. Cases have been found where some plants turned out to be good for alleviating some diseases, but the side effects could slowly harm people without knowing it. As correlations are insufficient, we should be concerned with understanding the origin and causes of what is happening in the real world. Once causes are found, knowledge strengthens, allowing us to develop new technologies. Knowledge leads to innovations that benefit the pharmaceutical industry, developing drugs whose side effects are better known in advance as they are better controlled.

In superorganisms, mechanisms like positive feedback keep the group together to achieve common purposes. Positive feedback can enable cultural trends, ideologies, generalization, information spread in the network, the adoption of technologies, and the eventual emergence of complex innovations. The reliance of our knowledge on correlated data is high. This knowledge may be transmitted to other generations of superorganisms using an assortment of complex languages. Nowadays, these languages can come as programming languages, pictograms, hieroglyphs, illustrations, and ordinary and sign languages. From creating and refining a complex language to constructing artificial machines, humans forge the path to reducing environmental uncertainty.

To protect the members of our superorganisms from random dangerous situations generated by nature, like other superorganisms, hurricanes, extreme heat waves, pandemics, famine, or climate change, we must be able to formulate causal models that allow us to forecast the future dynamics of them all. Physical theories and mathematical models use the present and the past to provide us with what is necessary to design weather forecast models using supercomputers and massive databanks, air conditioning to keep us cool, accurate meteorological analysis to understand climate change, or the development of food technologies to avoid famine.

Scientific superorganisms put a lot of effort into providing us with the best possible forecast. Everything that we can predict will allow us to differentiate between what is purely random and what contains regular patterns and coherences. If we follow the behavior of these patterns, we can better predict how they will behave in the future.

***

By nature, we try to get what we can out of the few regularities that are grasped by our brains. Unfortunately, there are cases where we cannot understand what is going on in a given situation. Scientists say the problem is like a "black box" when this happens. A black box defines systems or events whose internal dynamics are unknown, but their outcome can be perceived. We can build a theoretical model based on how this black box interacts with the outer world. Since we do not know what happens inside the black box, we create an 'artificial' theoretical model that explains what is possibly happening. Typically, this artificial model will be supported by a set of patterns (logical explanations or mathematical equations) that we have developed to support our conclusions. For example, it is known that we all live in a world described by three-dimensional space (length, width, and depth) (Figure 11). However, defining a macro-world with three dimensions does not allow us to understand what is happening in the micro-world. Theoretical physicists have found that to explain the micro-world (particle physics), we may need to incorporate more spatial dimensions than we can perceive. This is how String theory defined a world with twelve dimensions that might not exist (Who can tell so far?).

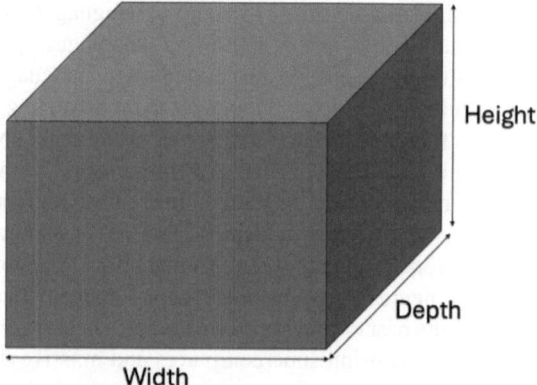

Figure 11: Three-dimensional object/space.

Contrary to what it seems, the microscopic world is not the only scenario where artifacts we cannot naturally observe have been devised to add completeness to theoretical models. The macroscopic world can also be described with more than three dimensions to explain counterintuitive phenomena. Chaos theory[cxxvii] relies on mathematical models that consider more than three spatial dimensions. This theory tries to explain why specific chaotic systems behave as they do. This is why, if we attempt to mathematically model the trajectory of an unknown chaotic system (black box) that evolves, we must add more dimensions to the model if the trajectory crosses itself at some point when we plot it. This latest step is mandatory if we want our model to be consistent with the theorems that support differential equations theory (the trajectory of a given system described by a differential equation is unique and, therefore, cannot intersect with itself[cxxviii]).

We could also develop a theory that explains the same phenomenon with the number of spatial dimensions we can naturally observe (three dimensions: width, depth, and height). But this is a risky aspiration since we would have to develop a theory that explains the phenomenon and does not contradict the theories that we already know explain other related phenomena. The risk is relatively high. We feel safer if we continue to build more layers of knowledge on top of what is already known. Many things could happen on that adventurous road. However, in many cases, scientists (Einstein, Hawking, Pasteur) have challenged all that was known, and this is how science and technology advanced.

Perhaps there are more spatial dimensions than the three we can perceive. But after all, spatial measurements are concepts that we have also invented. The concept of dimensions describes spatial objects in our macroscopic world. If incorporating more features in physical theories can help explain many phenomena, then there should be no problem. After all, we were left in this universe with little or no prior knowledge of what is going on around us. So, understandings or pattern configurations we grasp will always be helpful as long as we can prove that (i) they model reality, and (ii) we can predict future developments.

The Uncertain

Humans often rationalize from data. We create a model of the world where we link present and past events to generate a temporary explanation of what is happening. These models come with an associated uncertainty derived from the impossibility of grasping everything. Whatever errors we get when forecasting the future based on our models is due to uncertainty. Linked to possibilities that cannot be predicted, uncertainty shows up like an information void. However, we should not be discouraged by that. In fact, among the diversity of options, the certainty lies.

In the absence of certainties, diversity is needed to help us find the best route. A narrow range of possible but diverse solutions to a problem is fundamental when considering the best path to solving it. Problem-solving is an ability that depends not only on the model of the world one can have; it is also contingent upon a collection of skills reinforced throughout a lifespan. The reinforcement of skills requires training to develop the right level of competency that aids life navigation. Each skill is a building block that can interact with others to contribute to tackling complex projects. Without clear knowledge of the ideal skill needed to accomplish a goal, a diversity of skills is recommended. For astronauts to go into space, they must accumulate at least 1,000 hours of pilot-in-command time on jet aircrafts[cxxix]. It is believed that flight experience helps develop and strengthen practical skills for astronauts and improves decision-making that will lead to survival in space. The exact skill set required to accomplish this Pharaonic job is still unknown. Nevertheless, the exact skill set is embedded in the handful of diverse skills that astronauts need to learn before being sent to outer space.

Diversity helps reduce prediction errors. The object of interest is placed at an intersection among the possibilities unfolded from diverse experiences, skills, knowledge, data, intuitions, or beliefs. Diversity supports brain training based on exposures. The astronaut case differs because the number of experiences going to outer space on a rocket is more limited than popular activities on Earth. On the other hand, sailing is a trendy activity on the Ontario Lake in Canada that can be easily accessible by many compared to outer space activities. A sailing boat apprentice who has never experienced extreme winds can struggle to adjust the sails that control the boat. If the apprentice has enough opportunities to deal with various tides and wind speeds, they can develop a concrete, valuable skill set in extreme cases.

The benefits of not limiting ourselves to a single skill set but extending its range are ample. As complex problems arise, solutions may be heterogeneously mixed with multiple knowledge and experiences beyond initial intuition. Personalities like Leonardo da Vinci—a painter, sculptor, architect, engineer, and scientist—was an authentic polymath with highly developed problem-solving capabilities. Gottfried Wilhelm Leibniz—mathematician, physicist, philosopher, and diplomat—invented calculus while writing works about politics, music, games, history, law, theology, etc. Motivated people like that acquire a massive repertoire of skills and knowledge useful for contributing to knowledge or advancing sciences. The benefit that multidisciplinary activities bring to the table is that they allow for elements from different sources to interact in the mind. This interaction can result in little sparks of creativity and innovation. This is also the case for the success of the acclaimed mathematician and Fields medal winner Ngô Bảo Châu, whose achievements are coupled with his ability to master many different subfields within the field of mathematics[cxxx]. The combination of many abilities can give birth to impressive and imaginative results.

Currently, the internet and the media are contributing to the emergence of more diversity of information, leading to more enriched thinking. People connected to the internet are stimulated to interact more with each other through what appears to be an infinity of channels. This has contributed to the rapid development of digital societies by democratizing information modes. At least in Western societies, people do not have to belong to a very privileged social class to access knowledge in digital format. However, the resources to implement our knowledge and skills in practice may be hampered by our circumstances, geography, society, and government. Accessing resources that feed and enhance our skill sets through something as personal as a computer or phone can lead to new creative solutions.

As resources gain accessibility, interactions between scientific and technological worldwide communities improve. A direct benefit is a sense of increasing certainty among ordinary people regarding complex matters. The globalization of knowledge through the internet lets us have access to the same information from any time zone, online doctor appointments and 24/7 healthcare for non-critical assessments, real-time access to bank statements and management of money, accurate weather and air quality indices updates, worldwide conferences, tracked food and goods delivery, books and scientific articles, cinema, 3D interactive museums, and archeological sites, high-resolution real-time world map view, video chats, AI-based assistants, etc.

Diversity and globalization also come with demons. As information becomes more accessible, a few superorganisms monopolize the platforms through which access to information is enabled, consequently establishing new privileges that will continue to belong only to a few. Human behavior has been analyzed, categorized, and monitored by monopolies like social networks that filter and standardize how people interact and what type of information reaches them based on their profiles. Despite the openness of digital platforms, exclusivity is still a label for clustering people and information differently. Access to scientific papers under subscription, getting accepted to be part of private digital clubs or organizations, or signing legal agreements of non-disclosure to access privileged information demark a line of exclusivity rooted in all societies one way or another. Structural dominance by digital platforms results from the homogenization and standardization of what are considered correct actions at the international level, regardless of the context in which they occur.

Dominance relationships inevitably emerge through these superorganisms artificially emerging from global human interactions. Michael Foucault had already predicted and explained this fact with mastery[cxxxi]. He argued that members of the Western society interact with each other based on relations of power. This theory shows that master-slave dynamics can define human interactions, where the person who exercises power (master) forces others (enslaved people) to follow them. The enslaved person cannot exercise the same control over the master. This type of unidirectional dynamic constitutes the nucleus of unequal social hierarchies. The unidirectionality in the interactions between social layers may be considered a source of the stability of a given society with this underlying structure.

The lower layers may receive directions from the upper layers but not vice versa. As in a pyramid, base layers do not have enough autonomy to change the shape of the whole structure—since they do not see the extent of it from the lower side. The pyramid may restructure and become something else if they gain enough autonomy. Changing the tip of a pyramid is more accessible than changing the base layers, which require more energy and risk irreversible structural damage. Bidirectional hierarchies are possible, where the positions of master and enslaved persons are exchangeable. More equalitarian functionalities may become apparent. The homogenization of power is a complex base for an organization beyond our linear conception of it. After the agricultural revolution, many societies are accustomed to a pyramidal scheme of power and control of communities and resources (Figure 12). In a unidirectional society led by a monarchy (pyramid tip), a commoner (pyramid base) cannot become a king because they were not born into the right family. A fundamental structural change in that pyramid is expected for that commoner to get there. In an attempt to obtain a bidirectional hierarchy with the emergence of democracy, privileges based on family lanes are mighty but not mandatory for commoners to access power. After the power has been consolidated, the equality could break, and a pyramid may materialize once again.

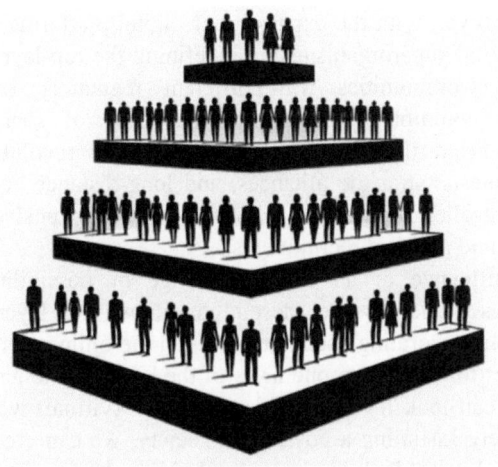

Figure 12: Social stratification pyramid.

Social and Environmental Order

The structures of societies depend on how connections between members are arranged—their level of organization. The most disorganized social structures tend to be seen as less thoughtful and more "left to chance" in many respects. Although disorganized, these structures achieve weak stability and are prone to structural damage by internal or external sources. Due to such weaknesses, their dynamics are prone to collapse, leading to catastrophic avalanches of unexpected events. As elements of the pyramidal layers go back and forth from top to bottom quickly, structural stability and composition cannot reach any stable equilibrium. A new form of organization could eventually dominate a weak network if an equilibrium is self-achieved or imposed. If the imposition of the new order is unnatural for the network, it will lead to catastrophes. If it reaches a degree of stationary stability, it will remain propped up with the strict constraints that allowed its establishment.

On a planet with more than 8 billion inhabitants by 2024, there is more possibility for the emergence of a large diversity of superorganisms. Data released by those is increasingly challenging to decipher. Shaped by top-layer structural frameworks like social networks, laws, policies, regulations, education, religion, culture, media, etc., we can see how concepts like personal autonomy and freedom are highly debatable. Without constraints outlined by those frameworks, an organization is not ensured. However, with the expansion of digital platforms worldwide, emergent digital superorganisms are redefining the top-layer structural frameworks. Communities with different regulatory backgrounds converge in common pools despite the reach of social network regulations. From this platform, they may plan revolutions, riots, boycotts, protests, strategic alliances, and long-distance actions using remotely-controlled drones and robots or achieve extensive religious, educational, and political recruitments.

The confluence of an extensive range of possibilities at our fingertips has made us freer despite traditional top-layer structural regulations. Incorporating social networks in the online world gives us more chances to jump from one layer of the hierarchy to another more quickly. We can look for our position in a new (virtual) world instead of immediately assuming a position in society. We can create avatars, enhance our physical look in pictures and videos, look more intellectual and knowledgeable, or belong to exclusive international communities. In short, we can constantly reinvent ourselves. This world has added more dimensions to our reality; it enriches our social experiences (for better or worse). Our social experiences are evolving every day simply because there are newer things we can access (both natural and human-created). Our ambitions do not stop at the digital level with the rapid advance of AI, hyper-immersive experiences like the metaverse, or remote manipulation of the market. We keep on building and destroying our physical reality. We look for expansion as the opportunities show up.

As mentioned previously, it is well known that humans have accelerated global warming, irreversibly affecting nature. All these changes increase the uncertainty of our environment, since it is constantly being disturbed by the extension of our ambitions. Such disturbances have consequences we cannot anticipate since our knowledge of nature is still limited. The world is changing and will keep doing so faster if we do not add the proper constraints to our collective developments. At some point, the world will change so fast that all the knowledge we acquire about it will become outdated before we have the chance to realize it.

The importance of trying to stop global warming is that this will prevent us (among other things) from starting from scratch. If we started from scratch, we would have to move to another planet or reconstruct this one. As nature is fundamentally complex and its behavior seems to be nonlinear, we will not be able to know a priori how the new order would be and how nature will behave. Chance will lead us to an undetermined transient time with unknown results. A different order is always an structure that is not known.

Regardless of the appearance of a new arrangement, randomness will always be stalking us through diversity. Although diversity improves our decision-making in most cases, this is still an open door to the unknown. Whenever something new comes our way, we must learn about it and incorporate it as a layer in our organization. Every time we have to build new knowledge, we start from what we know and begin to build a bridge towards the unknown, enroute for a new cloudy island whose shape we can barely see. Every time such a bridge is to be built, the risks and uncertainty increase to overwhelm us. But once we master this new knowledge, we grow and move on.

On Complex Oscillators

Simply put, our lives depend on the collective dynamics generated by the concatenation of several biological oscillators. Focusing on the sustainers of life, the circadian, respiratory, hormonal, cardiac, and digestive systems have associated rhythms related to their oscillatory properties[cxxxii]. The circadian oscillator has a sleep-awake cycle with an average period of approximately 24 hours. The respiratory system controls oxygen absorption and carbon dioxide discharge through breathing, with an associated inhale-exhale cycle period of around 3 seconds in a resting state[cxxxiii]. The rhythms of hormonal cycles lead to the regulation of multiple bodily functions. A remarkable hormonal process in women is primarily identified by periodic vaginal bleeding, with a cycle length of approximately 28 days. An average resting heart period ranges from 0.6 to 1 second per beat, which sustains consistent blood circulation throughout the body. The digestive process is less of a structural oscillator like other systems described above. However, excretion can take 24 to 72 hours, depending on the time of food ingestion. If food ingestion has a periodic pattern, excretion will do as well.

The synchronized neural activity in the brain also exhibits oscillatory behavior with associated oscillation periods. Almost 100 years ago, Hans Berger measured the first brain waves using electrodes on volunteers' scalps. The first formal publication on electroencephalography (EEG) was published by this German psychiatrist in 1929[cxxxiv]. The alpha waves measured in those experiments had a period of around 125 to 90 milliseconds when the subjects were resting in a quiet space with closed eyes. This regular oscillation was disrupted by various cognitive tasks and even by a simple eye-opening.

In later works, Berger identified the existence of beta waves, which have considerably shorter periods, reaching up to 8 milliseconds, associated with mental alertness. Nowadays, other brainwaves have been identified as delta, theta, and gamma, with ranges that may differ from what Berger initially defined. For instance, the gamma wave period ranges from 23 to 11 milliseconds, and beta waves range from around 76 to 33 milliseconds[cxxxv]. As these oscillations reveal brain function states, neuroscientists have appropriately categorized brain patterns throughout the decades.

All these biological oscillators are possible because their internal constituents are clustered in networks to carry out synchronized work. The staggering complexity of these networks is currently beyond our understanding. However, we are closer to getting better brain scans using new neuroimaging techniques. Regardless, such an interdisciplinary field like neurosciences has led scientists from other fields to intervene in the search for plausible explanations for the swarming complexity of the human brain. In a more abstract framework, neurons are interpreted as mathematical objects with associated network functionality. Depending on the degree of depth adopted by the scientists who develop these mathematical models, each object can range from being a simple variable, a black box defined with more than one variable, or a biologically realistic representation of a neuron in small or large networks. Variables are mathematical symbols that can be varied if some conditions are met. For instance, in Newton's second law of motion, the relationship between force and acceleration of an object is given by the equation Force = Mass × Acceleration. In this equation, we find three variables: Force, Mass, and Acceleration.

Experts in these fields usually start with the appropriate description of simple networks and then add more realistic layers to study the complexity levels. Complex systems are those composed of many objects that interact in some way. The more objects we can count in this context, the more complexity there is. Complex systems have the peculiarity of being more than the sum of their parts. When a set of elements in a network interact, they generate dynamics that cannot emerge from their constituent parts alone. We can state that the structure behaves like an organism. Hence, we cannot simply reduce an organism to a simple gathering of elements exchanging information. The number of elements and how they interact will define such an organism's complexity.

Interactions between elements are an essential part of the definition of an organism. When we talk about interactions, we must think about them as an exchange. What is exchanged can be heat, sound, energy, vibrations, particles, and more. In short, interactions are channels for transmitting information that allow data exchange. Elements can have one or multiple ways of interacting with each other. For example, between humans, we can find various types of interactions restricted by our limited sensory channels. This limitation may seem like a barrier, but it does not hinder the possible richness of interactions thanks to the number of people one person can interact with in a lifetime. Therefore, the amount of data transmitted through these channels is immeasurable, in addition to the incredibly sophisticated information processing machinery that deals with incoming data. Not only is the diversity of material received from the environment outstanding, but the complexity is compounded by the fact that each person can interpret it differently.

Complexity

We can receive information from multiple sources simultaneously through the channels that enable interactions in our bodies, such as the sensory system. On the street, we can receive auditory information from an immeasurable number of sources. There seems to be no limit to the number of stimuli we receive from the exterior. However, the limits are marked by our attention to such stimuli. Without the proper filters, we would be saturated with messages obtained through our senses from the outside world. Filters are even more necessary when the net interaction between a person and the environment seems irregular, almost random. If the stimuli are more synchronized, they can usually be perceived simultaneously.

Added to the complexity of our internal interactions, elements in our bodies (cells, organs, or systems) can get disturbed by environmental elements, such as viruses, bacteria, parasites, fungi, allergens, toxins, extreme temperatures, radiation, pollution, etc. To understand this better, breaking down our structures and their forms of interconnectivity can reveal what these external elements are doing to our bodies. This frame of reference takes us from the macroscopic to the microscopic world, allowing us to disentangle the chain of events that evolve from simple interactions to extended infections or diseases.

This deconstructive approach is useful as soon as we have a map of the internal composition of a body, before the interrelationships with external agents start. Constructing a map of an organism is complicated. For a map of any member of our species to be created, it would have to contain accurate connections from the micro to the macro. For this experiment to be successful, even what we believe is the most insignificant element that constitutes us should be included in the map.

If a particular element is not included, the map will be incomplete and represent something else. The resolution of this map should be high enough to accommodate all aspects and all interactions.

To better visualize what is behind this thought experiment, let's imagine that we have two copies of the same person: the original person and a reconstructed body following the guidance of a map. Suppose the copy was reorganized with an amino acid sequence missing an A (Alanine). Since the amino acid sequence determines the unique structure and function of each protein in the body, at least the role of one of the proteins in that body is compromised. This implies that some of the body's functions will differ slightly from what they would be in the original body. Given the non-linear nature of hierarchies that make up our bodies, it is not unreasonable to think that even that tiny difference between copy and original will become amplified over time. This will cause the two humans to behave differently after a while, as we usually observe in experiments with twins.

Our current technology does not allow us to map our bodies accurately. We are a long way from reaching that point. Making an exact map of our organisms is an individual process that must be carried out per person. Within our organisms, we interact with the different types of atoms, molecules, cells, organs, and systems that make us up. In particular, we find various information channels in the multiple configurations of brains, nervous, and digestive systems, making each of us unique, highly complex entities.

Figure 13: DNA illustration. Genome sequencing.

Two significant international efforts have begun with a view to a deeper understanding of the human body: The Human Genome Project and the Human Brain Project (HBP). The Human Genome Project (HGP) targeted the mapping of all genes of human beings, i.e., the genome (Figure 13). This ambitious project was launched in 1990 and ended in 2003, with impressive achievements after a 3-billion-dollar investment. The publication of a detailed description of the human genome was the most significant milestone, developing a reference genome containing around 92% of the total sequence. The reference sequence for the human genome is a model that serves as a base for comparing genetic variations in each person. With 8% gaps in the genome sequence by 2003, the sequence was highly accurate, but incomplete. In early 2022, a research article led by the National Human Genome Research Institute in the USA claimed to have achieved "The complete sequence of a human genome[cxxxvi]." This breakthrough was possible thanks to the development of better tools and techniques for preparing, mapping, and sequencing DNA[cxxxvii], which included advanced lab equipment and mathematical, computational, and statistical models.

The HBP was carried out between 2013 and 2023[cxxxviii]. With a budget of over 600 million Euros, this EU-funded flagship project aimed to advance our understanding of the human brain. The impressive digital reconstruction of the circuit of neurons in 3D was accomplished. The largest 3D map contains at least 200 brain regions, showing a static connectome of the brain. Despite the impressive results, this macro-project still needs to achieve its primary goal: mapping the human brain 100%[cxxxix]. Compared to the HGP, the HBP could not reach a comprehensive understanding of the brain at a cellular level in the first place. One of the reasons behind these results is the fragmentation of the project organizers, which limited collaboration among them. Beyond the critiques, both HGP and HBP have shown how considerable international efforts and a big chunk of investment can advance our worldview significantly. Nevertheless, getting a comprehensive reference model of a complex organization like the human body is an even more significant challenge.

***

The concept of complexity is widely used to describe large, tangled configurations. From what is comprehensible but hard to explain in a few words to whatever is beyond our understanding, multiple objects and events can fall within the limits of what the concept of 'complex' could entail. The broader the spectrum of elements that can be described with this concept, the more difficult it is to find a general, universal understanding of what complexity is. Generally speaking, complex systems comprise multiple variables (or elements) interacting somehow[cxl]. Complexity is measured by the number of elements in a structure and how such interactions are constituted. For instance, the interaction between one person and a thousand marine stones is not the same as between that person and a thousand others. Marine stones cannot provide feedback similar to people when an interaction is prompted. Modeling interactions between a person and three static marine stones can be simpler than between that person and three other people.

Due to the wide range of elements and interactions in complex systems, the joint dynamics that the network generates can range from regular to highly irregular. When uneven, these dynamics can be so bumpy that they resemble noise. As we have already learned, the key to knowing the degree of complexity of a system is not in excess elements but in how the components are structured and interact. For example, in chaos theory, it is well known that nonlinear systems comprising three independent variables can show chaotic behavior. In a few words, for a system to exhibit complex dynamics like chaos, we sometimes need three independent elements that interact irregularly with each other. These dynamics over time do not show a pattern understandable to humans, so we call them chaotic or quasi-noisy.

Although chaotic, some systems can appear regular and irregular depending on how their behavior is qualitatively represented. If we observe the temporal evolution of the famous chaotic Lorenz attractor, we will find this system behaving erratically (Figure 14). The Lorenz attractor was developed by Edward N. Lorenz in 1963 (Figure 15), describing the collective behavior of three variables (x, y, z) included in a simplified atmospheric model[cxli]. There, x represents the temperature differences between ascending and descending air masses, y is associated with horizontal temperature variation, and z describes the temperature changes with altitude within the fluid.

The behavior of individual variables $x, y,$ and z in the Lorenz interacting model is erratic and, therefore, unpredictable at first glance. But if we look at these three elements plotted together (known as the state space), we notice a beautiful butterfly-shaped figure (see Figure 15). This shape proves we have found a form of order in disorder. This example shows that the search for meaning is latent in all circumstances. The brain looks for patterns that satisfy our models of the world, but it cannot do this alone, so we have developed advanced techniques to get there. We required computers to solve the nonlinear differential equations that model the system to find regularities in the chaotic Lorenz dynamics. Before the age of computers, finding solutions to these equations systems was deemed very challenging. Finding order in tangled configurations is a task that may require advanced resources. Throughout our history as a species, our internal information processors have jointly developed analytical tools that will help us understand complex systems.

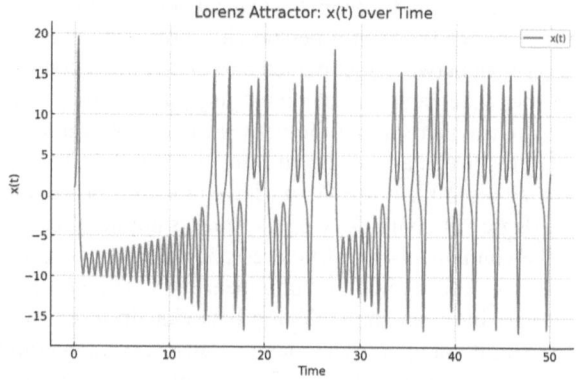

Figure 14: Lorenz attractor prediction of temperature differences.

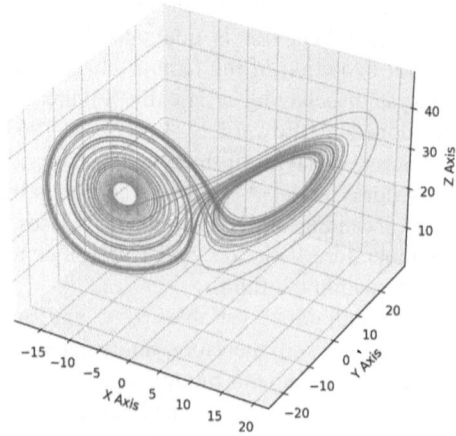

Figure 15: Lorenz attractor.

***

Often, to understand very complex systems, we need to simplify them to the point that they become a toy model. In the modern day, it is well known that we can design artificial neural networks that help us find forms of order in complex systems. For example, recently, Google DeepMind created a program called AlphaFold[cxlii], which can accurately predict protein structures from their amino-acid sequence. In this case, a reasonably complex system like a millions-parameter artificial neural network had to be designed to understand the structuring mechanism of another complex system called protein.

Even today, what these artificial neural networks do to solve challenging tasks has yet to have a single, definitive, and universal explanation. For now, it is inexplicable that these networks generate an expected response to a given stimulus once trained. The resulting trained artificial networks are highly complex, although less complicated than biological brains. This lack of complexity is related to the simplicity of the artificial neuron models we use to recreate brain functions. Artificial neural networks can be designed to encompass many more neurons than the average human brain. Still, more than a large number of artificial neurons is needed for a better brain model. The functional diversity of synaptic interconnects and the features of biological neurons are yet to be uncovered before we can entertain achieving a truly biologically realistic performance. We may be able to create AI models that surpass human performances, but this does not mean that these models can resemble superior human intelligence. The super-intelligence of those AI models may be fundamentally different from a person with an IQ above 140.

Usually, a system's complexity level can be measured based on various statistical and dynamic factors. For example, we measure repetitions in the behavior of complex systems to establish the probability that a given event will be repeated in the future. To do this, we let such systems generate dynamics for a long time until we can record as much data as possible until we find behaviors that turn out to be similar in time (invariants). Let us recall that invariants repeatedly appear with almost no change, indicating that they are nearly certain to happen again over time. By finding underlying (invariant) patterns, we may reduce uncertainty and increase the predictability of our system.

When a system is too complex, with excess elements interacting with each other in a tangled way, it is more difficult to find such regularities (or invariants) in its behavior over time. Perhaps such repetitive or invariant behaviors are there before our eyes, but we cannot elucidate them because our minds are mainly unprepared to detect such regularities. We can increase the predictability of this system if we reduce the number of elements and interactions that compose it. The increase in predictability is often related to lowering the given problem's complexity, mainly due to the reduction of elements that do not contribute to describing the system's behavior. This step is known as dimensionality reduction, and it is a valuable step to unmask hidden regularities in a mass of noise caused by elements that are not part of the core system. Denoising data is the first step when trying to lower the complexity of a system.

Oscillators

All known macroscopic entities can originally be studied from the reference framework of the complex systems theory. Whenever several elements interact through different channels, we can use the methods of the complexity theory to analyze the dynamics generated by the said elements. Many complex systems are studied as single entities with constituent elements that share similar features. Elements can be a group of neurons, bacteria, ants, mammals, superorganisms, etc. The common characteristic of all these entities is that they have internal rhythms to generate dynamics independently. This self-sustainability feature is the fingerprint of autonomous oscillators, which are the masters of changes in their behavior as time passes[cxliii].

Oscillators are entities whose behavior consists of cyclical fluctuations around an average value. Top and bottom fluctuations respect to the average are usually capped at values that depend on the amount of energy that the entity has. Such fluctuations or oscillations can be regular (periodic) or chaotic, depending on the internal composition of the oscillator[cxliv]. When two or more oscillators are coupled, the coupling connections will perform as communication channels. Depending on how robust the transmission channel is, the exchanged information throughout that channel may modify the elements' natural rhythm (internal dynamics) and eventually lead them to reach a collective rhythm. Such collaborative dynamics may have all kinds of behaviors ranging from regular to chaotic. The resulting dynamics are so erratic at times that the oscillators may get inactivated, a cataclysm known as oscillation death. When the collective behaves regularly or periodically, we expect all (or most) elements to follow a cyclical pattern. When discussing a cyclical tendency, we refer to elements with a behavior that seems almost repetitive when they develop activities together.

When we observe that the evolution in time of a complex system is entirely periodic, we define it as a perfectly regular system whose behavior will be repeated over time without expecting any change by itself. This means that every past behavior will be replicated in the future. This allows us to predict precisely what will happen in the future if no external agent disturbs that system. Complex systems capable of generating such dynamics have constituent parts evolving in complete regular agreement, or at least their evolution is simple for us to understand.

But as soon as the behavior of a complex system looks more erratic, these elements may not be as organized as those that behave periodically. Suppose, the evolution of such a system is so irregular that we cannot predict even slightly how it will act in the future. In that case, we define the system as behaving hyper-chaotically or randomly. But how can such an irregular dynamics manifest? Could all the elements have decided to oscillate in total disagreement? It is hard to give a general answer to this question. Conceivably, internal interactions are set so that components are disturbed by each other's very different rhythms, generating a collective response that is incomprehensible to us. For example, we see collective periodic dynamics in the spectacular Chinese military parades where everybody seems to be in sync, following each other without stepping in each other's way (Figure 16). Hyper-chaotic or random

Figure 16: Illustration of a synchronized military parade.

collective behaviors can be observed in violent riots or crowd crushes. It is certainly possible that there is a hidden organization happening within a riot, but an external viewer would likely not decipher it.

The overall dynamics of a complex system do not necessarily indicate that all its elements have independent individual behaviors. There are cases in which some elements decide to cluster and have their sub-collective dynamics. This means we can find several component subsets within a common rhythm. If several of these subsets behave differently, the dynamics of the whole collective appear to be somewhat irregular. This grouping property without consensus is common in complex systems with many elements.

As discussed in previous chapters, focusing on small samples of regularity in structures that behave randomly may be a mistake since these patterns could be generated by chance. There is a tendency to think there is no room for repetition, clustering, or some order in total randomness. Counterintuitively, random regularities are also part of chance. Sometimes, we cling so tightly to these insignificant random patterns that we believe there might not be randomness after all. These beliefs come from a biological survival strategy of a tireless search for patterns. It is important to note that when there are no consistent sources of certainties, we make them up by generalizing from those random forms of order we can find by accident. This is how certain philosophers and thinkers have rationalized the existence of faith in religion, politics, and conspiracy theories. Once a faith/belief is established, thinking that reality could differ can be challenging. Instead of changing their perspective, many can build rational theories or ideologies around their faith to incorporate them into a causal source of certainties.

***

Some complex systems are intuitively defined as 'complex' for a good reason: arranging them into a more manageable configuration where their behaviors are not erratic and unpredictable is difficult. We constantly search for the correct information to understand and twist these entities as we please. To get some insight into these structures, scientists have developed mathematical tools to analyze how erratic the behavior of a complex system is. One important assessment tool is the Lyapunov exponent, which determines how chaotic a given system is. This index assesses the rate of divergence in time between two trajectories that were initially very close to each other in an oscillator. If these two trajectories diverge or separate entirely at some point by themselves, we can state that the underlying system is chaotic—due to sensitivity to initial conditions[cxlv]. In systems with regular behavior, if two trajectories start their way together, they should keep that trend forever unless an external disturbance causes the divergence or the system runs out of energy. The more variables the complex, nonlinear structure has, the more likely it is to behave hyper-chaotically or almost randomly[cxlvi].

If a complex system has a random behavior (like white noise), we cannot define how many variables generate such complexity. We cannot create a deterministic mathematical model of it. The number of contributing elements is predefined as infinite (i.e., uncountable, immeasurable), so we cannot expect to find reliable patterns by chance, just as we cannot cling to reliable patterns in life. We hope to see less unpredictable behaviors in systems with far fewer elements (which we can study analytically with our mathematical tools). But in general, oscillators are usually composed of many components—this is especially true if we consider subatomic compositions in the mix. Nevertheless, organic oscillators can exhibit synchronized behavior despite such a high degree of complexity when coupled. In what follows, we will see how complex structures reach certain agreements, which result in synchronization.

Collective Dynamics

Agreeing to cooperate or to follow others' paths is one of the reasons humans have achieved so much. Phenomena that involve communication between two or more complex entities until they agree to behave similarly have been widely studied in the past few decades. This agreement that results in the emergence of a collective rhythm is known as synchronization. It is a widespread phenomenon in the natural and artificial world. Many structures with different levels of complexity turn out to be 'synchronizable' if they are adequately coupled to others akin. Synchronization is one of the most wonderful natural generators of consistent order in our complex reality. This marvel is founded on the fact that numerous very complex structures can be synchronized with their peers, which is quite impressive[cxlvii].

More formally, synchronization is achieved when two or more bodies with oscillatory behavior interacting with each other reach an underlying agreement and begin to develop dynamics whose rhythms (or natural frequencies) are indistinguishable. Although there are many types of synchronization in dynamical systems theory, this description attempts to show the most intuitive way to understand the definition of synchronization. Living and artificial entities with some cyclical temporal (oscillatory) evolution are prone to being synchronized by others akin. This characteristic that living and artificial organisms have in common gives us the feeling that synchronization is a universal phenomenon.

An oscillating body (an oscillator) can exhibit periodic, quasi-periodic, or chaotic behavior depending on its constitution. As discussed in the previous chapter, we can interpret the human body as an organism capable of behaving as an oscillator, which can be almost periodic at times. Our respiratory, cardiac, circadian, hormonal, blinking, walking, and running motions are examples of oscillatory behavior. As we will deduce from the upcoming explanations in this chapter, our bodies result from the coupling of many internal oscillators inhabiting us.

As oscillators, we are prone to synchronization if our bodies interact with others. To understand this phenomenon, imagine we are in the gym exercising with a stationary bike. Around us, we only hear other people cycling (oscillating movement) on their bicycles, and from time to time, we turn and perceive their movements or feel the vibrations generated by their rhythms through the floor. And so, without realizing it, we suddenly begin to follow the cycling rhythm that our neighbors show. We then realize that we have unconsciously synchronized ourselves with the rhythm of others. This happens quasi-automatically, or at least without premeditation, almost like naturally.

Synchronization is a natural, stable equilibrium state reached by two or more interacting oscillating bodies. If some conditions are followed. Due to the exchange of information via interactions, these bodies persistently disturb each other. Typically, punctual disturbances are identified as foreign, destabilizing the receiver. However, if the bodies are synchronized, neither body feels a disturbance as such. So why do two or more bodies decide to oscillate at the same pace despite the persistent pushes that take them on and off equilibrium? If the body's rhythms (natural frequencies) are not too dissimilar, or their communication channels are not too weak or too strong, those bodies may eventually synchronize. When bodies are synchronized, the disturbances cease as they become one entity. Since the disturbance has oscillations similar to those of the disturbed element, they will not catastrophically push each other out of their equilibrium states. Consequently, they can reach a shared state of equipoise.

***

Often, synchronization is used to describe phenomena unrelated to body-to-body synchronism, but to another phenomenon known as 'resonance'. In physics, resonance is a phenomenon that occurs between two or more interacting oscillators. However, before these begin to interact, one of the oscillators is in an active state (it oscillates), and the other oscillator is in an inoperative state (it does not oscillate). Once the oscillator interacts with the non-oscillating object, it forces the latter to change its state to active, i.e., the object begins to oscillate. As in synchronization, the two bodies oscillate with the same frequency. It looks as if the driven oscillator exhibits an exaggerated behavior after being disturbed by the external oscillator forces.

The difference between synchronization and resonance is that when the interaction ends, the object being somehow 'forced' to oscillate stops doing so. In contrast, in synchronization, both objects will continue oscillating despite not interacting. Synchronization involves adjusting rhythms between two or more bodies exhibiting oscillating dynamics. Resonance is a body's response influenced by external forcing, where the driver's characteristic frequency matches the driven body's natural frequency.

With these concepts in mind, we can understand the underlying dynamics behind social interactions, such as social resonance[cxlviii]. In group dynamics, the forcing oscillating object (the leader) starts interacting with the others through speech. This speech is delivered with a natural frequency (charisma or natural rhythm of the speaker) capable of influencing many people. The agents being influenced (people) are like the objects forced to oscillate or to act in our analogy with physics. In reality, the driver can be interpreted as a particular idea that does not feel unnatural and highly influences the mindset of receptors. It is as if people received a shot of energy in the neural circuit representing that idea in their minds. Some might relate this information to the famous phrase "this idea resonates with my thoughts"—indeed, it might be like that.

It is important to note that the characteristic frequency of the forcing agent (the leader) must be in tune with the individual natural frequencies of the agents that resonate with it (people). Hence, charisma is vital because it is a dynamic with an associated rhythm that no one could elude. It gives the feeling of mutual understanding in tune with people's needs. Excess charisma is dangerous because it mesmerizes and seduces people to follow the leader unthinkingly wherever they go. This is how populism has become popular lately in Western societies. Leaders can get people to follow them through the particular frequency of their voices and the rhythms of their speech. When those rhythms are the right ones, an automatic excitation is generated in the followers. This situation is more common than it seems in group dynamics. We see how many people resonate with their leaders' ideals (or rhythms).

In situations of actual synchronization, people would synchronize with the actions that leaders would already have been doing. Leaders would not have to use gimmicks to make people follow them—their actions should be enough to make people go along with them. Highlighting these differences is essential since it shows that following leaders does not come from a consensual agreement (synchronization) but from a forcing action (resonance). Although both phenomena (synchronization and resonance) are equally natural, synchronization is less invasive and more balanced for individuals in the community.

***

The resonance between two coupled oscillatory systems, A and B, occurs when B begins to oscillate, and this happens when the frequency of the oscillations of A coincides with those of B. It is through resonance that radio telecommunications operate. For example, if A and B are antennas, the transmission of information (effective communication) occurs when A (transmitter) and B (receiver) have the same frequency. In the case of synchronization, the systems involved would already oscillate with a very similar natural frequency (although not necessarily the same) when they are not coupled.

If we analyze the type of interaction between two oscillators more closely, we realize that they have an excitatory and inhibitory interplay. Excitatory and inhibitory dynamics cause bodies to regulate each other through feedback until they agree. The excitatory dynamics of one body are inhibited by the excitatory dynamics of the other and vice versa. Synchronization will occur more quickly and efficiently when the interacting oscillators are identical and exhibit a reasonably regular (at least almost periodic) behavior. Such oscillators become a single complex oscillator when synchronized.

Speaking of complex systems that can be synchronized, we can find excitatory-inhibitory dynamics that leads to oscillatory behavior in humans. It cannot be said that every dynamics that groups of people generates can be interpreted as the result of oscillations, but many things fall into this category. Many human oscillating features allow the existence of the synchronization phenomenon when the conditions are given. The example that was introduced earlier about synchronizing people in the gym makes sense in this theoretical framework since cycling is an oscillatory activity that can allow for synchronization if there is interaction between several cyclists through sight, or even vibrations on the floor.

It may eventually be noticed that humans somehow briefly synchronize very frequently with each other. The more complex the oscillatory task we carry out, the more difficult it is to maintain this synchronization in time[cxlix]. Highly complex structures behaving irregularly have more difficulties when attempting to synchronize and maintaining such synchronization for long periods. On the other hand, much less complex systems (such as two pendulums interacting through the ceiling that supports them) do much better in this because they cannot behave hyper-chaotically or quasi-randomly on their own.

We should be grateful that 'rules' have emerged in that way because life itself could not be possible if the synchronization between simpler systems was not robust enough. As discussed in other chapters, cells in the heart need to be in sync for the heart to function correctly. When the synchronization is broken, in this case, the heart begins to behave erratically in an arrhythmic way, so it is advisable to include a pacemaker to help restore the synchronization of these cells. A neglected arrhythmic heart is more likely to stop working than others, so in this case, the synchronization break would destroy a system on which many other organic subsystems depend. Life itself is generated from well-defined patterns formed by elements (cells) interact and synchronize with each other.

It is hard to imagine what would become of us without synchronization in many body parts. Studies have shown that even long-term memory is linked to robust phase synchronization between neurons, allowing stored information to stay there longer[cl]. Long-term memory has contributed to our intellectual advancement as a species, along with synchronization between people to achieve common goals and constitute superorganisms.

An Agreement to Light the Way

In the brain, timing plays other equally important roles. From a purely individual point of view, each brain is designed and strengthened through training to find regularities efficiently. Consequently, a lot of the regularities we uncover (or pay the most attention to) will partially depend on how or what we were trained for. If the training of a brain depends on environmental conditions, the brain will strengthen synaptic connections that enhance task performance in those conditions. It will weaken those that correspond to less priority situations.

Strengthened synaptic connections last over time, and the information they retain becomes memory content. Thus, when the brain receives details about elements already existing in memory, specific neurons will be activated, and others will not. This activation indicates that the brain recognizes, at least in part, the facts that are entering. When the information is of hybrid origin, several regions of the brain may be activated in a synchronized way. As researchers French and Thomas[cli] discuss in their paper *The Dynamical Hypothesis in Cognitive Science*, the representation of a red car is not made by the same neuron in a single part of the brain; it is done by firing two different clusters of neurons, where one corresponds to the concept of red and the other to the car. By synchronization, we mean that both clusters may fire at the same time.

This form of synchronization between brain regions allows more efficient information processing. Other forms of information representation are found in the Mirror Neuron theory which explains how the activation of specific neurons due to an action perceived in another person coincides with an action we have performed in the past. A team of neuroscientists from the Universita' di Parma in Italy have shown that particular neurons in a monkey's brain are activated once it watches another monkey (or a human) perform the same action. This action can be as simple as spitting or scratching. When the monkey under study repeats the same actions he has seen in others, the same neurons fire.

Synchronization may get triggered between the times the action is perceived and the activation of the neurons involved. This manifested regularity shows our brains' efficiency when analyzing information registered in memory. Our internal information processor does not have to explore everything from outside from scratch; it only has to represent it by activating neurons already trained to process such data.

From this point, we can extrapolate this knowledge to more complex tasks, such as conversations between two people. Research led by a team of scientists from the Basque Center on Cognition Brain and Language in Spain has shown that the brains of two people talking are somehow synchronized. Suppose two people have to analyze data during a conversation. In that case, the most efficient thing is that they must have similar representations of the information they exchange in their brains simultaneously[clii]. In this way, these people can understand and follow each other. Such an interaction promotes synchronization between movements, postures, articulation, and tone as the conversation develops. Processes of this type increase communication efficiency in terms of energy consumption to process what enters and leaves their brains and speed of response.

Given the efficiency associated with the synchronization that occurs when information is processed, it is worth wondering if neural synchronization is related to intelligence. To try to answer this question, we refer to a study led by scientists from the MIT in which evidence was found that brain wave synchronization enables rapid learning[cliii]. Brain waves are characterized by the oscillations produced by neuronal activity—neuronal oscillations. Therefore, if several neuron clusters are activated simultaneously in different brain areas and synchronize their oscillatory rhythms, new communication circuits are enabled more efficiently. Such learning efficiency has been identified as 'correlated' with general intelligence in other studies. However, this approach has been quite controversial since the learning process can be slow or fast depending on the learning strategies implemented. Inadequate learning strategies can lead to total failure, as in some school systems that restrict the development of a child's intellectual activities rather than promote them.

As developed in this section, regular patterns generated by synchronization help establish complex structures such as human beings (their brains, their internal organs, and even the society in which they live). But is synchronization the only feature necessary for order to exist? Possibly not! Another additional set of processes can intervene in forming organized structures. Hierarchies would be behind many of these compositional phenomena that have proven robust and essential for the subsistence of many living and even artificial structures.

Complex Waves

The behavior of synchronized objects is quite peculiar. The coupled elements involved in total synchronization behave like a wave, like a mass of objects with the same spatiotemporal evolution type. In contrast, coupled components that are not synchronized can be perceived as individual particles moving in all directions while being disturbed by external agents. These waves can also expand in size when their constituent segments are synchronized with more elements of the same type. If the elements connected have a different nature (natural rhythm), the resulting hyper-structure may not synchronize, but this does not mean it cannot reach any organization. Then, synchronization is one type of organization that may exist.

A complex system may become a hierarchy when several structures with different rhythms are coupled to form an organized hyper-structure. At this point, all the constituent substructures create a single complex body with a higher-order shape. Suppose, we still see that there are differentiable groups in the hierarchy. In that case, these are composed of clusters of elements that behave consistently even though they are coupled to other bodies that may disturb them.

Due to its superior organization form, the total dynamics resulting from the hierarchy also take the form of a wave (as a single structure). Any microscopic change in these dynamics is so tiny that it cannot be wholly differentiated but contributes to the maintenance of the wave. This happens because hierarchies are composed of levels, where the highest ones dictate the direction in which the entire structure will move. The members of the lower levels can move freely but always within limits imposed by the higher levels. This is why the hierarchy's behavior does not change despite its movement. It holds up despite small changes from its lower levels. These structures exist thanks to the balance achieved by constituent elements. Such a state of equilibrium can be natural (through synchronization) or imposed through force.

The organization of hierarchies has the same purpose as any other type of formation: to reach a stable state of equilibrium in which the intrinsic nature of the system is maintained despite small changes within it. Otherwise, the hierarchy could be broken since the lower excess of freedom would influence it. Some hierarchies can be affected by the movements of their lower levels. In these, the higher levels may change due to variations in the lower levels. This interplay yields to new forms of stability that alternate—never staying in one. Cases like that occur when the structure of a hierarchy is dispersive and nonlinear. Here, small changes at the lower levels propagate through the structure, leading to different forms of organization in the hierarchy.

The success behind the organization of hierarchies is that the members of lower levels typically exert less influence on other elements (they have fewer connections). Hence, they are sub-structures with a low level of complexity. Because of this, lower levels are less likely to modify the actions of higher levels. The number of connections in a hierarchy increases as higher levels appear. It is said that the highest level may be the most complex and has the most connections because it exerts influence on all levels—it can regulate and reorganize the hierarchy. Its dynamics spread through the entire hyper-structure or superorganism and affect all levels.

Changes generated at a lower level might not modify higher levels if the hierarchy has a linear constitution. In nonlinear hierarchies, connections are irregular and monotonically growing downward (from highest to lowest level). Here, the connections could be scattered, and the lower layers could interact directly with the higher ones somehow; this is how lower-layer changes affect higher layers as well.

This is how these superorganisms take the form of an organ, in which the system behaves as a single entity and cannot be understood if it is decomposed into parts. When disconnecting elements and trying to analyze them individually, it is often difficult to know why the organ behaves this way. We must constantly refer to the global dynamic's context for a better understanding. For example, we would have to analyze the actions exerted on each element when interacting with others. The context is what could unite the organ (or holistic structure) when trying to study it in parts in a reductionist way. But even so, it takes work to establish the connection between the reductionist and holistic perspectives.

A clear example of this is found when we analyze artificial neural networks. In principle, networks made up of Perceptrons are straightforward to describe[cliv]. They are compact mathematical models that encompass the most general functions of biological neurons. Synaptic weights determine the connecting force between the perceptrons. The central operation of a perceptron can be described as a multiplication between the input information and the synaptic weights. Then, the result of each multiplication is added, and an activation function is assigned to transform this result into a threshold output. Such is how we obtain the artificial neuron (perceptron) response to an input stimulus (information) (Figure 17).

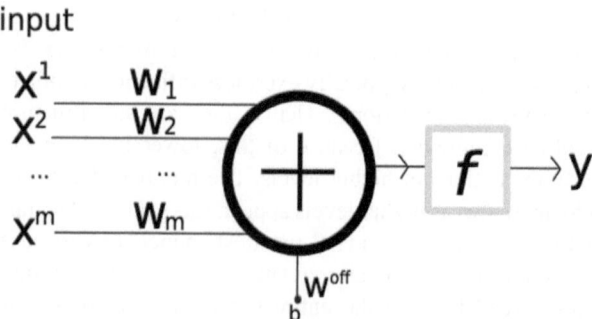

Figure 17: Neurons receive information via inputs X transmitted through their connections (synapses). The strength of each connection is indicated by its weight (W). The neuron's cell body sums these weighted inputs and transforms them using an activation function f. W^{off} are weights for bias b.

Notably, an artificial neuron's action is simple and can be described in three steps (multiply, add, and activate) in a reductionist framework. We might expect a neural network composed of these elements to be easier to understand than biological neurons, but this is different. Researchers in AI still need to develop a universal model to understand what these networks do to process information as efficiently as they do. For many, artificial networks are still an enigma and a black box. Many advances try to explain their operation, but more is needed. One answer to this is that when perceptrons are coupled in a neural network, they behave like an organ, and their action can no longer be individually understood.

Progress has also been made in studies of brain function through EEG analysis. Little by little, neuroscientists decipher the specific neuronal responses associated with our actions. They are trying to identify and label them. However, we need a causal model to understand the organ from its constituent elements and their interactions.

***

We find hierarchies everywhere in human communities. Comfortable or not with hierarchies as the optimal form of organization, it is widely spread throughout the planet. Most successful companies of our time are shaped as hierarchies, where the highest levels make important decisions related to their evolution and future. However, they are dominated by the internal needs of the state or the laws of the country in which they are located. In the context of the world's nations, the state decides what rights and duties citizens should have according to their needs. In certain countries, people's rights could be considered deplorable from the point of view of others, and from the point of view of the former, the latter's rights could be seen as excessive. For some authoritarian states, too many personal rights could be considered dangerous as the hierarchy becomes more vulnerable to radical changes.

How is society defined in terms of a hierarchy[clv]? As we can see, society is unlikely a perfectly linear structure where the lowest levels can never access the highest. History has shown that social hierarchies can collapse and give way to new formations (in which members of the lower levels could now be part of the higher layers), as happened during the French Revolution and before Napoleon took over power as an emperor. Then, it is valid to ask how it is possible to maintain societies under such social stratification if the lower levels can access the higher layers with a low probability of success.

In our times, we live in the era of information globalization. Somehow, we discover what other states and societies do and their decisions. Even certain states cluster and make decisions together. Consequently, members of the highest layers of different hierarchies are incorporated into the highest layer of a superorganism, for example, the United Nations, the European Union, the Organization of American States, etc. Through establishing a much larger hierarchy (hyper-hierarchy), where the distance between low and high levels is very long, changes pushed from the lower levels may need to be more evident. To counteract this supremacy, massive protests are organized by ordinary people when controls exerted over them are unbearable.

No system lives in isolation on this planet; if lower layers can eventually access higher layers, it is because connections can be naturally made. Members of the bottom layers can start their hierarchies in the digital world. Exchanges on non-regulated platforms in the Deep Web can quite disturb physical organizations. Digitally created hierarchies can force the modification of many aspects of traditional hyper-hierarchy organizations. Related to this, the so-called terrorist organization ISIS managed to spread to many parts of the world, and its consequences affected millions of people belonging to all social levels[clvi]. For example, if not for the Internet, such information would have been more restricted to the Middle East.

Information is a power[clvii]. Information behaves like a packet of energy that spreads through the levels of a nonlinear hierarchy and can modify or destroy it. This energy can settle in the lower levels of hierarchies and cause them to collapse from within. The history of humanity has seen how the lower levels of a society realize their underprivileged rights when compare themselves to others around through available communication channels (e.g. trade roads, gossip, internet, etc.). This information transfer is implemented in the form of adequate reactionary energy that the oppressed use to destabilize the structure they belong to and thus change the constraints the higher levels have imposed to them.

At this point, the information that spreads from one hierarchy (sender) to another (receiver) can be interpreted as a disturbance applied to the receiving hierarchy; when the ranking changes in the receiver hierarchy, new constraints, and level assignments can occur. This indicates that amid the disorder, lower layers could rise. Cases like that happen when a group of communities influence each other to change the established order through revolutions. Such a phenomenon may occur as the result of a chain effect.

To conclude, thanks to state-of-the-art media, even members of low levels can be part of other hierarchies and belong to high levels there, as we find in social networks. These people may influence what other hierarchies should do and how they should evolve. Globalization of online and offline interactions is making our societies much more complex, weakening traditional hierarchies through information trade. However, conventional structures always try to maintain the order that is shaking more and more. It cannot be said that we are on the road to anarchy (where high levels could disappear), but hierarchies certainly seem more vulnerable now.

Maintaining Order

How to keep order in what seems to be a world of unpredictability. Humans need help to maintain what we understand as order since our inertia often makes adapting to other forms of order hard. It is understood that we live in a world with immeasurable forms of order, and our mental capacities are designed to see only a few. This is not a problem if we live in a world isolated from external disturbances, and the form of order to which we are accustomed can continue undisturbed, and we with it. This utopia is persistently confronted with changes reality comes with.

Categorization is one of the most used organizational strategies in hierarchical societies. In the organization of a technology company, we have the CEO, Chief Scientist, CTO, COO, VP, directors, managers, coordinators, technical staff, HR, IT, janitors, etc. The bigger the company, the more categories it will have, and the more labels will be created to identify the role of each member of these categories.

When assembling models of hierarchies to make sense of their structure, we try to incorporate the disturbances that we receive from the outside into those models. Once a competing company appears on the market, it is quickly categorized as such. This means the competitor takes an intangible place in the model of the company's structure as a foreign element with potentially disturbing effects. If the interaction with an external element (e.g., a competitor) is not integrated into the models of a hierarchy, then, that external agent is considered a source of noise—a disturbance that is not assessed.

In an organized system, noise can have different effects. This can shake the hierarchy, or it can completely disintegrate it. This is one of the reasons why labeling and categorization are critical to maintaining order. What is not identifiable through a category or a label can be a source of noise that can lead to catastrophe. By developing structural models, we can incorporate the disturbances the superorganisms receive internally and externally. By combining such elements into the analytical model of a hierarchy, we assign a specific location in that model for the external agent. The position in the model will be defined by the number of active connections between the original hierarchy and the external agent—connections that it uses to disturb aspects of the system we know. Such an assignment increases the predictability regarding the future behavior of that system.

Behind the need to create mathematical or analytical models of organisms, there is an intense necessity to find causal relationships between individual components. By assigning places to external agents in analytical models of an organism, we increase the complexity of such an organism. Counterintuitively, by increasing such complexity, the predictability is also increased by proxy. Integrating the causes of an organism's erratic behavior can make us anticipate what will happen next. What was initially perceived as noise is now part of a more extensive and comprehensive model. Therefore, actions can be taken to balance such movements effectively.

No matter what, actual hierarchies have a high share of unpredictability due to (i) the impossibility of developing an accurate analytical model due to their inherent non-linear structures, and (ii) the hidden interactions with the outside. Unfortunately, there are an incalculable number of agents whose causal formulations are elusive. What is not understood is still interpreted as noise in the best-case scenario, or it can also be seen as an immaterial force that comes from beyond—and we may need to develop rituals to counteract it. In the face of true chance, many of us cannot deal with it as it is. A causal explanation must be behind it—even if the cause itself cannot be explained by material means. Without formal causal explanation, people can think that token sources of erratic outcomes can be spirits, negative vibes/energies, unbalanced chakras, god's punishment, the evil eye, demons, the devil, goblins, and fairies. Depending on the classification, various methodologies can be put in place to neutralize their effects, such as praying, crystals, incense, rituals, and ceremonies.

***

In maintaining order, the struggle between different interacting superorganisms has moments of standard order and reciprocal perturbation, creating imbalances. The standard order can come in the form of control that one structure exerts over another, such as master-slave forcing, or it can be an agreement between them, the result of some synchronization.

The history of relations between France and Germany can serve as an illustration for this argument[clviii]. These relationships have always had different shades depending on their degree of interaction. In the times of Nazi Germany, we saw how this country imposed a strong master-slave interaction between both nations, which led to a complete destabilization of the internal order of France. This historical incident resulted in Germany exerting substantial control over France during the Second World War, which attempted to hamper France's independence. Thereafter, dynamics between these two countries were self-regulated. At present, Germany and France have established communication channels through the superorganism known as the European Union. This collective agreement may still interfere with France's autonomy, strictly speaking. Nevertheless, as far as we can tell, a strong imposition is not currently the case.

This superorganism has changed over time as it self-regulates. Countries like England, with an assigned position in the hierarchy of interactions, have decided to disengage. Although important, England was not fundamental for such a structure's survival since it had been able to self-organize after the BREXIT[clix]. These spontaneous ruptures of some elements are not a fatal problem (at least in the short term) for well-constituted hierarchies. But is the survival of this organization the result of a robust structure? Or can chance sustain it? Certainly not for long.

Next Steps

The high complexity of all the structures surrounding us constantly challenges our internal information processing systems. Facing so much complexity with limited resources, human beings are always searching for more accurate answers to infinite questions regarding this world. We work to improve our understanding of the natural and artificial phenomena that surround us since our quality of life can continually improve, in turn, through knowledge. And even the survival of our species is attached to this search for knowledge and the push for innovation. In the future, we may not survive climate change or the breakdown of the sun. We may need to escape and colonize other planets of galaxies. Regardless of our ambitions and how they affect the Earth through nuclear devastation or excess GSG pollution, the sun may begin to die in billions of years. If we are not extinct yet, we must have enough technology to escape.

To accomplish such an overwhelmingly massive project, we need more. Ever since Homo Sapiens emerged and matured between glacier and interglacial periods, we have always dealt with environmental uncertainty. These cycles were accompanied by highly irregular climate and landscape changes that impacted generations of hominids. The imprecisions of such environments made us naturally inclined to reduce that uncertainty one way or another. In the last 7 million years, our brains have increased in size, providing more infrastructure to tackle uncertainty. More recently, we hope to find the long-awaited finite-time regularities through sciences and rigorous mathematical models. These regularities found through inductive inference methods are part of the kind of 'order' our brains understand best. But it may be in our best interest to know enough so that all our conclusions can emerge from deductions instead.

To improve our understanding by establishing certainties, we developed artificial tools that help us get there. Sometimes, we take inspiration from nature to create the artificial tools we need. Thus, we see how robotic arms are currently used for the mass production of commercial products, where each product has a high probability of being functional despite the speed with which they are produced. More complex robots like the Cheetah or the Boston Dynamics Atlas could be sent into space to explore new planets without risking human lives[clx]. In short, the creation of complex artificial entities inspired by biological entities continues to amaze us.

Artificiality

As our species browses the world, it faces its own limitations. From the basics, making a fire was a medium through which we could digest food more efficiently. Breaking down complex proteins, fats, and carbohydrates was simplified when cooking meals before ingestion. Also, killing harmful bacteria present in raw food prevents us from going through sick-recovering processes. Learning to make a fire and use it as an instrument to improve our quality of life and increase our survival chances has been a significant advance in our evolutionary process. The first evidence of the instrumentalization of fire as a food co-processor dates back more than 700,000 years ago (the exact date is still a great debate topic). The making of fire evolved throughout generations, leading to the gas stoves we use today, with gas imported directly to our homes from remote locations.

The means of moving from one place to another also developed over time. Much has happened along the way, from simple walks to intercontinental travel. From the date Homo Sapiens left eastern Africa, it took our ancestors more than 50,000 years to travel to South America via the Bering Strait[clxi]. In the 21st century, it may take about 21 hours to travel from Ethiopia to Peru on a flight crossing the South Atlantic Ocean. Horses, camels, bicycles, cars, trains, ships, airplanes, helicopters, rockets, etc. are arm's-length extensions of our locomotor systems, allowing us to go places with various efficiency rates (Figure 18).

Brain co-processors have also existed for a long time. The Sumerians invented an abacus more than 5,000 years ago to support more complex arithmetic operations than neurotypical brains can handle. From any form of primitive computers like the Sumerian abacus for arithmetic or the Greek antihythera mechanism for predicting astronomical phenomena up to advanced supercomputers in our current era, we have made abysmal progress[clxii]. In the past, computers were mainly used to help with arithmetic and calculus operations. More recently, higher-level cognitive functions are represented in silicon-based processors with the infrastructure provided by AI on fast computers.

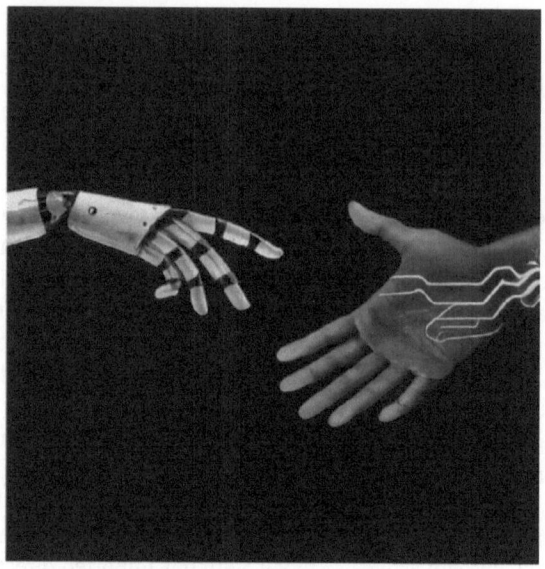

Figure 18: Future one-to-one collaboration.

Artificial brains are envisioned to be based on a simplified concept of neural networks[clxiii]. From what we can discern, the key to recreating a biological brain is the design of its neurons and how they interact. This step has proven challenging since we do not yet have a complete static reference of all neuron connections in a neurotypical brain, let alone a dynamic version. As with other examples mentioned above, the goal of AI may not be to create an accurate copy of the brain in silicon but a co-processor.

An extension and enhancement of brain functionalities should considerably amplify the scope of what we can achieve as a species. Researchers in AI have shown that neural networks can be designed to solve problems as complex as those solved by the brain, such as recognizing images or voices, translating texts from one language to multiple others, creating art, brainstorming solutions to given problems, summarizing a book, creating code from an algorithm, etc. These tasks can be achieved by training different neural network architectures that can solve those problems more efficiently, where neurons cluster in layers, and those layers interconnect in convenient ways[clxiv].

The vast majority of publicly available neural networks that solve the tasks described above are based on a generalization of the concept of a perceptron neuron. A perceptron can be defined as a mathematical model that encompasses some of the most essential functions of a biological neuron: multiply, add, and activate. Simplifying, the use of perceptrons assumes that a biological neuron receives information from the outside through its dendrites; this information is summed and non-linearly transformed by the neuron's body, after which we obtain responses in the terminal axon. The concept of perceptron simplifies the action of biological neurons in only three parts (which have proven to be quite relevant for processing information): input information to the neuron through input synaptic weights, sum weighted inputs, and applying an activation function. The activation function mimics the firing feature of biological neurons.

These networks alone cannot solve any problem; they must be trained to solve it. The most popular current method for training these perceptron-based networks is backpropagation[clxv]. In summary, network performance errors are reused to update the synaptic weights when training to solve a task. This process is carried out iteratively until the performance is the highest possible. Other methods, such as direct feedback alignment[clxvi] and equilibrium propagation[clxvii], are gaining much attention but have yet to be as widely used as backpropagation[clxviii].

Other neural models include spiking neurons, which attempt to mimic the behavior of biological neurons more accurately. In this model, the input and output of information are encoded in the time it takes for the neuron to fire again after the last time it was activated—the timing between spikes. Models that use spiking neurons are much less popular, as they are mathematically more complicated to train with traditional methods used by the AI community, such as backpropagation. Due to the extensive development and evolution of neural networks based on the concept of perceptron-based neurons, we will focus the rest of this book on this approach. In this way, we will review what the current AI community has achieved in the present and its prospects.

Silicon-based parallel processors are currently the core infrastructure supporting AI[clxix]. This hardware has enough parallelism to effectively represent billions of neurons exchanging information at high speeds. Such an architecture is closer to the functionally massive parallelism of the brain, which is linked to its efficiency when solving complex problems.

Since several circuits work simultaneously in different parts of the problem, the whole network behaves as several co-processors delivering chunks of digested data. The information exchange speed depends on the performance of transistors on a chip (Figure 19). These tiny elements are tuned to represent zeros and ones under some conditions. The binary property of these processors links AI to the field of digital computing, where the digits '0' and '1' are combined in multiple ways to represent portions of reality. Similar to the human brain, such representations allow for simulating physical reality in a different physical platform.

The speed of those transistors can reach GHz performance, meaning that the information exchange is in the nanosecond range, which is a billionth of a second. Compared to the millisecond (one-thousandth of a second) performance of synaptic interconnects, the interactions between transistors are fundamentally faster than biological synapses. Nevertheless, the vast parallelism of cognitive operations in the brain compensates for the lack of neuron-to-neuron speed. If the physical implementation of a realistic human brain comes true on a silicon chip, then, the information processing of such an artificial engine will beat humankind's performance at least a million times if we can compare them linearly[clxx].

Figure 19: CMOS chips for AI.

***

If our brain is as extraordinary as described in this book, why do we need co-processors for our fabulous brains? What is so challenging out there that cannot be addressed by us? So far, artificial networks are expected to serve all sectors of society. From video games to stock market prediction, artificial neural networks are slowly gaining ground in all possible areas. Since we are in the age of the internet, efficient telecommunications, and massive data banks, this is an excellent historical moment to start analyzing such massive datasets.

That collected data is often more than just a randomly arranged set of points. On many occasions, the data can reveal something about the structure that generated it. Whether data is from the weather, the stock market, or even chaotic traffic flows, we often perceive a particular underlying order. AI-based methods can help find hidden patterns in data from the examples described above. Based on an extensive weather dataset—that can be dozens of years long—neural network models are trained using a large portion of that dataset, and the remaining part is used for validation. When we have a dataset with data like the weather, we can split it into temporal chunks. The first chunk can contain data from 1988 to 2008—defined as a training set. The second portion includes data from 2009 to 2010—defined as a test set. AI-based models can be trained with the training set, and the model's accuracy can be tested using the test set.

Initially, sequential temporal data of this kind was processed with recurrent neural networks, where neurons are wired so that some (or all) of their output responses serve as input information to themselves. Then, each sequence element will enter the network in different time steps and remain stored in its internal memory by reusing such information through recurring connections. In this way, each neuron will remember the sequence element introduced in the previous time step. Such a type of network works well to analyze data that comes in a sequential presentation, such as generation and classification of text, translation, weather data, etc. Nowadays, variations of deep neural networks are at the forefront of data analysis and global forecasting[clxxi]. Deep networks do not necessarily have recurrent connections but can be designed to interconnect data differently while keeping strong forecasting capabilities.

Likewise, we can also do something similar with information from visual content. We would often like to organize a massive collection of millions of images. In reality, this tedious task could be performed by human agents, but it would take a long time and many motivated people due to the excessive volume of visual data. Tasks that require continuous attention are particularly relevant because they need a level of surveillance that is hard to achieve by ordinary people. Instead of having humans, security and surveillance systems could serve as real-time face classifiers of images captured using cameras. Unlike humans, these classifiers would stay energized and do their job 24/7 as long as electricity and maintenance were provided. Artificial organisms can do the job regardless of the motivation behind solving these tasks.

Some classification and surveillance tasks like that can be solved efficiently using convolutional neural networks. These networks filter the entire set of input images, one after another, and extract what is statistically familiar to them and what differentiates them. Then, they classify those images into different categories based on the filtering result. These networks are convenient not only for the security sector but also for the health sector. They can be beneficial for diagnosing diseases through images in the near future and support medical decision-making.

Other more straightforward types of neural networks help us reconstruct partially damaged images using a recurrent Hopfield-type neural network. Generative Adversarial Networks (GANs) can be the way to go for more advanced tasks. GANs have become quite popular because they generate realistic photographic images of human faces; they create photographic pictures of people that do not exist. GANs and convolutional neural networks are feedforward neural networks, where neurons are stacked in different layers without any recurrence between neurons, as is the case with recurrent networks. The AlphaGo Zero computer program—which learned how to play Go better than any human on the planet—is also based on a feedforward network. AlphaGo Zero has been designed using a residual neural network in which the outputs of neurons jump over layers to communicate with more distant layers. Another example of feedforward networks widely used today to solve complex problems is transformers, with which GPT-4 has been created. The GPT engine is a powerful text generator that tries to imitate how humans use language[clxxii]. Other architectures, like AlphaFold, can predict a protein's three-dimensional structure from an amino acid chain's folding and intramolecular bonding.

As well as these introduced examples, many other architectures solve all kinds of problems from low to high complexity. Shortly, more network models will appear to efficiently solve tasks that go beyond what some human cognitive functions can achieve. However, the question remains open as to whether what these networks are doing is solving problems intelligently (with artificial 'intelligence') or whether they are simply doing a very complex fitting of the data (which will not necessarily lead to the development of intelligence). To get closer to what intelligence encompasses, we will review its basic concepts once again in what follows.

On Artificial Intelligence and IQs

Counterintuitively, the most critical research and advances in AI are unrelated to the quest to generate general artificial intelligence per se. Instead, it is focused on creating intelligent artificial agents that solve particular problems beautifully. Research in AI is very focused on what is known as specific skills. In particular, the focus is on solving a distinct and well-defined problem, such as recognizing handwriting or translating texts. More complex tasks like deep reasoning and self-driving still need improvement.

Intelligence tests may be applied to determine if artificial and biologically intelligent agents can solve the task they were trained for. Different tests range from the most effortless (character recognition) to the most complex (recognition of human actions or reading comprehension), which can be applied to humans and machines. Generally, most tests of human special abilities are created so that a person finds solutions to a series of proposed problems for which they have yet to receive any previous training. In contrast, artificial agents need to be trained to do so. For example, an experienced driver can drive in different Western countries without requiring highly specialized training. Such a driver may only need a quick review of the manual and the rules—anything else required comes by intuition and extrapolation. A self-driving car is still struggling with the peculiarities of San Francisco's traffic.

One of the most significant limitations of AI is that it still requires some supervised training (via variations of the backpropagation algorithm). So, our artificial machines could not extrapolate their recently acquired knowledge to related but fundamentally different events. The type of artificial agents we have today continue to be trained to solve specific problems. Those artificial agents that solve many particular issues seem more 'intelligent' than others. Still, they are only trained to carry out those particular activities. For example, nowadays, we have AI technologies such as Apple Siri, Amazon Alexa, or Google Assistant, which are voice assistants trained to solve specific problems related to searches and answers to simple questions. These assistants cannot argue coherently and autonomously but support simple and specific daily tasks.

Currently, engines designed by OpenAI, Google, and Meta use a hybrid approach that pushes the bar to unimaginable positions. The generation of human-like scripts is based on supervised and unsupervised training[clxxiii]. Large language models are being trained in this process, which are behind the latest advances in artificially matching human cognitive functions. Researchers in AI use the internet as a data source for training their models to determine the next word in a given sentence. An exciting example of what these methods have achieved is ChatGPT from OpenAI. As for the version GPT4, the chat can interact respectfully with people from all over the globe. Through a method known as Reinforcement Learning from Human Feedback, filters have been added to avoid hatred and racist outcomes from the prompt. Despite the possibility of having advanced conversations with ChatGPT, it has yet to achieve truly human-like cognitive functionalities. However, the engine can perfectly work as a co-processor, supporting people in daily tasks.

***

To solve more complex tasks than "simple content searches", artificial assistants typically connect to remote servers to support their functionalities. For example, if a user wants to translate text using a cell phone, the translation may be done by an engine in a remote location. Hence, the translation software requires an internet connection to complete its task. This happens due to the need for more local computation power to solve such a task. For this reason, hardware companies are developing chips specialized in solving AI tasks efficiently. Those chips can be cointegrated into a cell phone or a computer to boost AI-related software. More functionalities can be embedded with hardware resembling humans' physical and cognitive structures.

In a very simplistic way, we can describe some of our cognitive processes as algorithms or a sequence of instructions. When translating text, making decisions, solving a problem, analyzing consequences, forecasting the future, etc., we usually have a process that can describe it more or less precisely. We can even rely on information from books, dictionaries, magazines, the internet, or people to complement our understanding of the problem to solve. In this era, it is not uncommon for people to find solutions on Google if they cannot get there by themselves. If the instructions are accurately described, we could sketch an algorithm that a computer can understand. However, most genuinely complex cognitive tasks cannot be represented with such a level of precision. We will soon realize that it is impossible to fill all the gaps in the algorithm when asked how we can make a difficult decision that may impact the rest of our lives, decide what is artisticly valuable, or any task that does not have a clear target even for us. Comparing human agents and artificial machines is still a very fuzzy enterprise.

So far, AI can mimic some functionalities that cannot be described with a list of instructions. However, they have yet to get the autonomy that humans have for consistent reasoning and consequent decision-making. Giving an artificial agent these abilities is quite tricky. If something similar is to be achieved with current supervised training methods, the AI would have to be trained to solve a practically infinite and random number of tasks by itself. Under this training, the machines cannot learn by themselves to solve problems for which they have yet to be trained. For this, the machine would have to be trained to learn by itself, and then it would have to create its filters or constraints to generalize from the experience correctly. In short, it will need to be able to achieve some level of awareness or consciousness.

Learning is adaptation. Learning is linked to understanding what is vital to reducing the uncertainty of the environment—what is necessary to survive autonomously. An autonomous system must self-organize to survive; it has to self-adjust to adapt to its environment and learn to solve problems posed by the circumstance. Its success will be based on how it has been trained to learn new skills that will lead it to survive. In theory, it is impossible to train a human or artificial agent for everything; it must be taught to learn to solve each problem that is presented to them. To do this, humans invest a lot of time and physical energy in their early development. Parents teach their children to solve problems independently so they can get enough autonomy for what comes when they are not around. However, they can still find other humans or resources to assist them. Getting an artificial processor to achieve this autonomy level is beyond our understanding.

Certain advances in this area already exist today. Many artificial agents that are used to play Go or generate text do not need specific training to carry out their functions step by step (Figure 20). The key to their success is that they can train themselves to solve a particular problem at a time. This field is advancing towards autonomy progressively more, but will it be enough to resemble human autonomy and intuition?

***

A remarkable characteristic of intelligence tests belonging to the field of psychometrics is the lack of absolute metrics. As discussed in Chapter 5, intelligence through the label of IQ relies significantly on the statistical methods used to shape the bell curve. With a necessity for getting samples that contribute to such a statistical pool, measuring IQs for artificial agents is limited. If an AI engine could solve all tests encompassed in standard general intelligence tests, we could position such an agent in a proper context. A highlight of these tests is that one cannot be trained to pass them. The stereotypical feature in IQ tests is the surprise since they measure problem-solving skills, adaptability, and efficient navigation of complex scenarios. An AI's degree of adaptability needs to be considered when assessing its intelligence level.

Examples of a specialized artificial engine without cognition are robots that serve caffe lattes or make cheeseburgers with great skill, but are dysfunctional when faced with variations of those activities. Being programmed to do a handful of things and being unable to diversify could be categorized on the low intelligence side of the curve. Even in the context of specific skills, not being able to learn to extrapolate is a struggle. From a caffe latte to a cappuccino, robotic hands must be reprogrammed entirely with new instructions. A person may also be trained to do so. However, with enough trial and error, a neurotypical person can develop a methodology to eventually figure out what would be required to make a cappuccino. The autonomy given by our cognitive apparatus facilitates adaptation even in the simplest tasks. The inability to reason beyond what an agent was trained to do limits its adaptability.

AIs have more cognitive complexity than simple robot arms. To begin, a machine can be (supervised or not) trained to play Go based on all known examples and strategies of games in the history of humankind. Once trained, such a machine is ready to generalize, but only in the field related to that game. This means that it can solve new problems once trained in that domain. This ability has resulted in AlphaGo winning over the best biological player in the world, Lee Sedol. One of the criticisms that could be made of this form of training is that the AlphaGo trainers probably used game examples from Lee Sedol. If so, the machine could have used Lee Sedol's intrinsic playing patterns to predict his moves, thus allowing it to plan counter-plays. This is an exceptional stepping stone towards adaptability since the AI uses what is known, adapts to the player's strategies, and proceeds accordingly.

Beyond the type of data with which AlphaGo was trained to beat Sedol, what is fundamental to understand is that behind games of this type, there is a finite set of simple rules that can be learned and never vary. Games are generally interesting because they are ideal (with unbreakable rules that never change and are known) despite their underlying randomness. In particular, games of this type are designed to have potentially infinite possibilities. It is likely to predict only some of these infinite scenarios a priori. Still, a player can try to predict what their opponent will do in each scenario with a certain associated probability.

Figure 20: Go board.

In the absence of concrete rules that we know are unbreakable, life is a game of imperfect information. The regulations and constraints of real life are often changing, adjustable, and interpretable. Extracting patterns that can guide our way is much more challenging. The most practical thing is to learn from experience and hope future scenarios resemble those we have already lived in. In this way, we could extrapolate knowledge. In the event of unfamiliar situations, relying on our adaptability is an asset. The adaptation process will lead us to find the rules that govern the phenomena behind new scenarios and adjust our response mechanisms.

Hyperspecialization

Even though general intelligence has been highly debated among experts, there still needs to be a solid consensus among all scientists in the field. For comparison, what IQ implies is not interpreted by all psychologists in the same way as physicists see in the Laws of Newton or the constant of light. No matter how complex, someone who excels at a single task can have an average IQ. Chess grandmasters have, on average, an IQ of around 100, which is the average IQ. The savant-like skills are not heavy enough to push the bar in their favor when using psychometrics for assessment. Despite their highly specialized abilities, grandmasters still manage to have enough cognitive infrastructure to navigate life without exceptional struggle.

From the looks of it, the simple fact of having extensive neural networks trained to solve some super-specialized tasks brings us closer to creating a true AGI, but more is needed. So far, we are expanding the range of tools to better solve a handful of problems. If we give a machine one of these specific tasks and nothing else, it can solve them with incredible precision—including the most complex games in the world.

When a similar protocol is offered to an average neurotypical person, they may need help to solve a complex problem efficiently and accurately. Nevertheless, exceptional human cases can perfectly compete with artificial machines in the efficiency/accuracy game. An example of this is the savants. People with the Savant syndrome can carry out the resolution of hyperspecialized tasks to the maximum. We found the case of Laurence Kim Peek (also the inspiration for *Rain Man*—the movie, 1988), who could memorize more than 6000 books on different topics such as history, music, literature, and geography, in addition to having developed skills to calculate calendars accurately and know the zip codes of all areas in the USA. Another example is Stephen Wiltshire, who can draw from memory entire cities that he has seen only once for less than an hour. These surprising abilities to carry out a limited number of tasks, as precise as a computer, are accompanied by other cognitive and motor deficiencies that hinder their everyday lives.

Studies on savants have shown that their highly developed skills are coupled with deficiencies in many other aspects, which may include difficulties in socialization and personal care, among others. For these reasons, savants were initially given the derogatory name "Idiot Savants"[clxxiv] to refer to their categorization of people with an IQ between

70–85[clxxv] but who could excel at specific tasks in an almost inhuman way—from the point of view of neurotypical people.

What bears the distinguishing mark between neurotypical people and others is precisely the unique way the neurons in their brains are interconnected. For example, Magnetic Resonance Imaging (MRI) studies on Laurence Kim Peek's brain have shown that the corpus callosum and other essential parts of the central nervous system were damaged. The corpus callosum is the brain region that divides the cerebral cortex lobes into the right and left hemispheres and, in turn, allows communication between both hemispheres. Since the function of the corpus callosum is to integrate motor, sensitive, and cognitive performances between the two hemispheres, if there is not correct communication between them, motor and mental problems can appear. For example, control of the fine movements of the face that give way to speech is limited, as well as being able to tie shoes or brush teeth.

Despite the muscular disability caused by such brain damage, these individuals have developed computational abilities beyond those imaginable by other human beings. Here, the key to success is developing a prodigious memory and a high capacity to focus on the task and absorb all it entails. Savants can quickly become engrossed in a specific problem until it is entirely resolved (as if motivation is maximized)[clxxvi]. However, only some of the existing tasks in the world fall within the domain of what a Savant can solve. According to psychiatrist Darold Treffert[clxxvii], a handful of categories of Savant interest can be listed: music, painting/sculpture, calendar calculator, mathematical calculation, and spatial skills such as accurate distance measurement and mapping, ease of learning multiple languages, navigation, etc.

Patterns emerge if we intersect all these categories and try to find commonalities. The excessive repetition of each controlled task may be one of them. Tasks that are less structured can challenge those systematizing minds. For instance, activities that include dealing with complex emotional interactions or touring a city like Cairo at 5 p.m. can take much work for these brain structures. The evidence indicates that environments that perpetually change can be challenging for savants. In these environments, the noise index exceeds the local predictability index; thus, finding durable and reliable patterns in such an environment seems almost impossible.

The intuition to understand specific patterns coming from emotions has many practical applications in our daily lives. Neurotypical people take it for granted that regularities they infer in their daily lives are necessary to function in highly uncertain environments. But for those people for whom these skills are absent, they are assigned the disability label so that they are supported to deal with these problems. The problem may be that most environments in our modern world have this high degree of uncertainty.

Besides the complicated abstraction previously raised, a Savant would still be deeply distanced from becoming an inspiration for current AIs in every sense. In their quest to adapt to the world around them, a Savant would learn how to solve the problems that obsess them, find the necessary data, and develop new tools that would allow them to excel in it (which requires creativity). Likewise, progress in AI will largely depend on some enhanced unsupervised training and the establishment of autonomy for learning and seeking knowledge. Otherwise, we will need to embody AI and give it a sensory system, a nervous system, and a stomach so it can develop subjectivity and awareness.

Creating a more predictable world for everyone is one way to pave the way for a functional general AI. Nevertheless, even if we incorporate many constraints and rules to follow (as in strict dictatorships such as North Korea or even in Margaret Atwood's dystopian novel *The Handmaid's Tale*), reality is still changing independently of how much we control it. However, as we all know, this utopic realization does not bring us closer to a safer world since we are still part of a universe independent of our decision-making and ambitions.

Bibliography and Further Reading

[i] Joan Stiles and Terry L. Jernigan. "The Basics of Brain Development." *Neuropsychology Review*, 20(4), (2010), 327–348. <https://doi.org/10.1007/s11065-010-9148-4>.

[ii] Bente Pakkenberg and Hans Jørgen G. Gundersen. "Neocortical Neuron Number in Humans: Effect of Sex and Age." *Journal of Comparative Neurology*, 384(2), (1997), 312–320. <https://doi.org/10.1002/(SICI)1096-9861(19970728)384:2<312::AID-CNE10>3.0.CO;2-K>.

[iii] Joan Stiles. *The Fundamentals of Brain Development.* (2008), Harvard University Press, *JSTOR.* <https://doi.org/10.2307/j.ctv1pncndb>.

[iv] Ken Ashwell. *The Brain Book: Development, Function, Disorder, Health.* Paperback, 2nd Edition (2019). Firefly Books Limited. <https://www.barnesandnoble.com/w/the-brain-book-ken-ashwell/1111981302> [accessed 16 February 2024].

[v] Rebecca Chamberlain and others. "Drawing on the Right Side of the Brain: A Voxel-Based Morphometry Analysis of Observational Drawing." *NeuroImage*, 96, (2014), 167–173. <https://doi.org/10.1016/j.neuroimage.2014.03.062>.

[vi] Emily S. Finn and others. "Functional Connectome Fingerprinting: Identifying Individuals Using Patterns of Brain Connectivity." *Nature Neuroscience*, 18(11), (2015), 1664–1671. <https://doi.org/10.1038/nn.4135>.

[vii] Sandra Ackerman. *Discovering the Brain* (Washington (DC): National Academies Press (US), (1992). <http://www.ncbi.nlm.nih.gov/books/NBK234151/> [accessed 16 February 2024].

[viii] Douglas J. Futuyma. *Evolution*, 2nd Edition (2009). Sinauer Associates Inc.

[ix] Stephen C. Cunnane and Michael A. Crawford (Eds). "Energetic and Nutritional Constraints on Infant Brain Development: Implications for Brain Expansion during Human Evolution." *In:* The Role of Freshwater and Marine Resources in the Evolution of the Human Diet, Brain and Behavior. *Journal of Human Evolution*, 77, (2014), 88–98. <https://doi.org/10.1016/j.jhevol.2014.05.001>.

[x] Manu S. Goyal and others. "Feeding the Brain and Nurturing the Mind: Linking Nutrition and the Gut Microbiota to Brain Development." *Proceedings of the National Academy of Sciences*, 112(46), (2015), 14105–14112. <https://doi.org/10.1073/pnas.1511465112>.

[xi] Pierre J. Magistretti and Igor Allaman. "A Cellular Perspective on Brain Energy Metabolism and Functional Imaging." *Neuron*, 86(4), (2015), 883–901. <https://doi.org/10.1016/j.neuron.2015.03.035>.

[xii] Alexa R. Romberg and Jenny R. Saffran. "Statistical Learning and Language Acquisition." *WIREs Cognitive Science*, 1(6), (2010), 906–914. <https://doi.org/10.1002/wcs.78>.

[xiii] Damasio, Antonio. *Feeling & Knowing: Making Minds Conscious.* (2022). Penguin Random House. 9781524747558: Books – Amazon.Ca <https://www.amazon.ca/Feeling-Knowing-Making-Minds-Conscious/dp/1524747556> [accessed 17 February 2024].

[xiv] David Eagleman. *The Brain: The Story of You* (2015). Vintage. <https://www.ebooks.com/en-ca/book/2074014/the-brain/david-eagleman/> [accessed 16 February 2024].

[xv] Helen M. Bramlett and W. Dalton Dietrich. "Long-Term Consequences of Traumatic Brain Injury: Current Status of Potential Mechanisms of Injury and Neurological Outcomes." *Journal of Neurotrauma*, 32(23), (2015), 1834–1848. <https://doi.org/10.1089/neu.2014.3352>.

[xvi] Joseph Jebelli. *How the Mind Changed: A Human History of Our Evolving Brain* (2022). Little, Brown Spark; Suzana Herculano-Houzel and others. "The Elephant Brain in Numbers." *Frontiers in Neuroanatomy*, 8, (2014), 46. <https://doi.org/10.3389/fnana.2014.00046>.

[xvii] Gerhard Roth and Andrea Schmidt. "The Nervous System of Plethodontid Salamanders: Insight into the Interplay between Genome, Organism, Behavior, and Ecology." *Herpetologica*, 49(2), (1993), 185–194.

[xviii] Gerhard Roth and Ursula Dicke. "Chapter 11: Origin and Evolution of Human Cognition." *In*: Michel A. Hofman (Ed.), *Progress in Brain Research, Evolution of the Human Brain: From Matter to Mind* (2019). Elsevier, CCL, 285–316. <https://doi.org/10.1016/bs.pbr.2019.02.004>.

[xix] Morgan Sheng, Bernardo L. Sabatini and Thomas C. Südhof. "Synapses and Alzheimer's Disease." *Cold Spring Harbor Perspectives in Biology*, 4(5), (2012), a005777. <https://doi.org/10.1101/cshperspect.a005777>.

[xx] P. Collins, Adam Martin and E. Squires. "Particle Physics and Cosmology." (1989). <https://www.semanticscholar.org/paper/Particle-Physics-and-Cosmology-Collins-Martin/3b2e26584c7e4b3afee207b4e2a151319ee0744a> [accessed 16 February 2024].

[xxi] Leon M. Lederman and Dick Teresi. *The God Particle: If the Universe Is the Answer, What Is the Question?* 1st Edition. (1993). Boston: Houghton Mifflin Co.

[xxii] Ralph Petrucci and others. *General Chemistry: Principles and Modern Applications*, 11th Edition. (2016). Pearson Canada.

[xxiii] Stanley L. Miller. "A Production of Amino Acids Under Possible Primitive Earth Conditions." *Science,* 117(3046), (1953), 528–529. <https://doi.org/10.1126/science.117.3046.528>.

[xxiv] Gerald F. Joyce and Jack W. Szostak. "Protocells and RNA Self-Replication." *Cold Spring Harbor Perspectives in Biology,* 10(9), (2018), a034801. <https://doi.org/10.1101/cshperspect.a034801>.

[xxv] Sohan Jheeta. "The Landscape of the Emergence of Life." *Life,* 7(2), (2017), 27. <https://doi.org/10.3390/life7020027>.

[xxvi] Jason P. Schrum, Ting F. Zhu and Jack W. Szostak. "The Origins of Cellular Life." *Cold Spring Harbor Perspectives in Biology,* 2(9), (2010), a002212. <https://doi.org/10.1101/cshperspect.a002212>.

[xxvii] Paul S. Cohen and Stephen M. Cohen. "Wöhler's Synthesis of Urea: How Do the Textbooks Report It?" *Journal of Chemical Education,* 73(9), (1996), 883. <https://doi.org/10.1021/ed073p883>.

[xxviii] Hugh G.M. Hill and Joseph A. Nuth. "The Catalytic Potential of Cosmic Dust: Implications for Prebiotic Chemistry in the Solar Nebula and Other Protoplanetary Systems." *Astrobiology,* 3(2), (2003), 291–304. <https://doi.org/10.1089/153110703769016389>.

[xxix] Kensuke Kurihara and others. "A Recursive Vesicle-Based Model Protocell with a Primitive Model Cell Cycle." *Nature Communications,* 6(1), (2015), 8352. <https://doi.org/10.1038/ncomms9352>.

[xxx] R. Ramachandran and C.D. McDaniel. "Parthenogenesis in Birds: A Review." *Reproduction,* 155(6), (2018), R245–257. <https://doi.org/10.1530/REP-17-0728>.

[xxxi] David B. Sauer and Da-Neng Wang. "Predicting the Optimal Growth Temperatures of Prokaryotes Using Only Genome Derived Features." *Bioinformatics,* 35(18), (2019), 3224–3231. <https://doi.org/10.1093/bioinformatics/btz059>; Johannes Rousk and others. "Soil Bacterial and Fungal Communities Across a pH Gradient in an Arable Soil." *The ISME Journal,* 4(10), (2010), 1340–1351. <https://doi.org/10.1038/ismej.2010.58>.

[xxxii] Michel Brunet and others. "A New Hominid from the Upper Miocene of Chad, Central Africa." *Nature,* 418 (2002), 145–151. <https://doi.org/10.1038/nature879>.

[xxxiii] J. Jebelli. *How the Mind Changed: A Human History of Our Evolving Brain.* (2022). Little, Brown Spark.

[xxxiv] L.P. Spear. "The Adolescent Brain and Age-Related Behavioral Manifestations." *Neuroscience & Biobehavioral Reviews,* 24(4), (2000), 417–463. <https://doi.org/10.1016/S0149-7634(00)00014-2>.

[xxxv] Charles A. Nelson. "The Development and Neural Bases of Face Recognition." *Infant and Child Development,* 10(1–2), (2001), 3–18. <https://doi.org/10.1002/icd.239>; Cara H. Cashon and Christopher A. DeNicola. "Is Perceptual Narrowing Too Narrow?" *Journal of Cognition and Development,* 12(2), (2011), 159–162. <https://doi.org/10.1080/15248372.2011.563483>.

xxxvi Kevin N. Laland and William Hoppitt. "Do Animals Have Culture?" *Evolutionary Anthropology: Issues, News, and Reviews*, 12(3), (2003), 150–159. <https://doi.org/10.1002/evan.10111>.

xxxvii Cristine H. Legare. "Cumulative Cultural Learning: Development and Diversity." *Proceedings of the National Academy of Sciences*, 114(30), (2017), 7877–7883. <https://doi.org/10.1073/pnas.1620743114>.

xxxviii C.M. Heyes. "Social Learning in Animals: Categories and Mechanisms." *Biological Reviews of the Cambridge Philosophical Society*, 69(2), (1994), 207–231. <https://doi.org/10.1111/j.1469-185x.1994.tb01506.x>.

xxxix Steven Pinker. *The Blank Slate: The Modern Denial of Human Nature.* Illustrated Edition (2003). London: Penguin Books.

xl Jason G. Fleischer. "Neural Correlates of Anticipation in Cerebellum, Basal Ganglia, and Hippocampus." *In:* Martin V. Butz and others (Eds.), *Anticipatory Behavior in Adaptive Learning Systems.* Lecture Notes in Computer Science (2007), 19–34. Berlin, Heidelberg: Springer. <https://doi.org/10.1007/978-3-540-74262-3_2>.

xli Goichiro Toyoda and others. "Electrocorticographic Correlates of Overt Articulation of 44 English Phonemes: Intracranial Recording in Children with Focal Epilepsy." *Clinical Neurophysiology: Official Journal of the International Federation of Clinical Neurophysiology*, 125(6), (2014), 1129–1137. <https://doi.org/10.1016/j.clinph.2013.11.008>.

xlii Mary B. Kennedy. "Synaptic Signaling in Learning and Memory." *Cold Spring Harbor Perspectives in Biology*, 8(2), (2013), a016824. <https://doi.org/10.1101/cshperspect.a016824>; Nima Mesgarani and others. "Phonetic Feature Encoding in Human Superior Temporal Gyrus." *Science*, 343(6174), (2014), 1006–1010. <https://doi.org/10.1126/science.1245994>.

xliii Usha Goswami. "Mind, Brain, and Literacy: Biomarkers as Usable Knowledge for Education." *Mind, Brain, and Education*, 3(3), (2009), 176–184. <https://doi.org/10.1111/j.1751-228X.2009.01068.x>.

xliv Laura Gwilliams. "How the Brain Composes Morphemes into Meaning." *Philosophical Transactions of the Royal Society of London. Series B, Biological Sciences*, 375(1791), (2020), 20190311. <https://doi.org/10.1098/rstb.2019.0311>.

xlv Daniel Swingley. "The Roots of the Early Vocabulary in Infants' Learning From Speech." *Current Directions in Psychological Science*, 17(5), (2008), 308–312. <https://doi.org/10.1111/j.1467-8721.2008.00596.x>.

xlvi Ariel M. Cohen-Goldberg and others. "The Interface between Morphology and Phonology: Exploring a Morpho-Phonological Deficit in Spoken Production." *Cognition*, 127(2), (2013), 270–286. <https://doi.org/10.1016/j.cognition.2013.01.004>.

xlvii Dale Purves and others. "Modification of Brain Circuits as a Result of Experience." *In: Neuroscience.* 2nd Edition. (2001). Sinauer Associates. <https://www.ncbi.nlm.nih.gov/books/NBK11005/> [accessed 17 February 2024].

xlviii E.F. Konrad Koerner. "The Sapir-Whorf Hypothesis: A Preliminary History and a Bibliographical Essay." *Journal of Linguistic Anthropology*, 2(2), (1992), 173–198.

xlix Jonathan Winawer and others. "Russian Blues Reveal Effects of Language on Color Discrimination." *Proceedings of the National Academy of Sciences*, 104(19), (2007), 7780–7785. <https://doi.org/10.1073/pnas.0701644104>.

l Matthew Sturm, Jon Holmgren, and Glen E. Liston. "A Seasonal Snow Cover Classification System for Local to Global Applications." *Journal of Climate*, 8(5), (1995), 1261–1283. <https://doi.org/10.1175/1520-0442(1995)008<1261:ASSCCS>2.0.CO;2>.

li Pertti Hella and others. "Disordered Semantic Activation in Disorganized Discourse in Schizophrenia: A New Pragma-Linguistic Tool for Structure and Meaning Reconstruction. *International Journal of Language & Communication Disorders*, 48(3), (2013), 320–328. <https://doi.org/10.1111/1460-6984.12011>.

lii Gemini Team and others. "Gemini: A Family of Highly Capable Multimodal Models." (2023). arXiv, <https://doi.org/10.48550/arXi

liii Edward Dolnick. *The Writing of the Gods: The Race to Decode the Rosetta Stone*. (2021). Simon and Schuster.

liv Susan Schaller and Oliver W. Sacks. *A Man Without Words.* (1995). Berkeley: University of California Press.

lv Lynne G. Duncan. "Language and Reading: The Role of Morpheme and Phoneme Awareness." *Current Developmental Disorders Reports*, 5(4), (2018), 226–234. <https://doi.org/10.1007/s40474-018-0153-2>.

lvi Josh Chartier and others. "Encoding of Articulatory Kinematic Trajectories in Human Speech Sensorimotor Cortex." *Neuron*, 98(5), (2018), 1042–1054.e4. <https://doi.org/10.1016/j.neuron.2018.04.031>.

lvii E. Kymissis and C.L. Poulson. "The History of Imitation in Learning Theory: The Language Acquisition Process." *Journal of the Experimental Analysis of Behavior*, 54(2), (1990), 113–127. <https://doi.org/10.1901/jeab.1990.54-113>.

lviii Barbara Tversky. *Mind in Motion: How Action Shapes Thought.* 1st Edition. (2019). Basic Books.

lix Danielle S. Bassett and others. "Learning-Induced Autonomy of Sensorimotor Systems." *Nature Neuroscience*, 18(5), (2015), 744–751. <https://doi.org/10.1038/nn.3993>.

lx Evelyn Tang and others. "Effective Learning is Accompanied by High-Dimensional and Efficient Representations of Neural Activity." *Nature Neuroscience*, 22(6), (2019), 1000–1009. <https://doi.org/10.1038/s41593-019-0400-9>.

lxi Moheb Costandi. *Neuroplasticity.* (2016). The MIT Press.

lxii D.J. Futuyma. *Evolution.* (2009). Sinauer Associates Inc.

lxiii Giuseppe Fusco and Alessandro Minelli. "Phenotypic Plasticity in Development and Evolution: Facts and Concepts." *Philosophical

Transactions of the Royal Society B: Biological Sciences, 365(1540), (2010), 547–556. <https://doi.org/10.1098/rstb.2009.0267>.

[lxiv] Lewis Dartnell. *Origins: How Earth's History Shaped Human History*, Illustrated Edition. (2019). New York: Basic Books.

[lxv] Larry R. Squire and others. "Memory Consolidation." *Cold Spring Harbor Perspectives in Biology*, 7(8), (2015), a021766. <https://doi.org/10.1101/cshperspect.a021766>.

[lxvi] Mariam Arain and others. "Maturation of the Adolescent Brain." *Neuropsychiatric Disease and Treatment*, 9, (2013), 449–461. <https://doi.org/10.2147/NDT.S39776>; Catherine A. Hartley and Leah H. Somerville. "The Neuroscience of Adolescent Decision-Making." *Current Opinion in Behavioral Sciences*, 5, (2015), 108–115. <https://doi.org/10.1016/j.cobeha.2015.09.004>.

[lxvii] Yingxu Wang and Guenther Ruhe. "The Cognitive Process of Decision Making." *International Journal of Cognitive Informatics and Natural Intelligence (IJCINI)*, 1(2), (2007), 73–85.

[lxviii] Christopher G. Coutlee and Scott A. Huettel. "The Functional Neuroanatomy of Decision Making: Prefrontal Control of Thought and Action." *Brain Research*, 1428C, (2012), 3–12. <https://doi.org/10.1016/j.brainres.2011.05.053>.

[lxix] Michael Guillen. *Bridges to Infinity: The Human Side to Mathematics* (1984). Los Angeles: Boston: J.P. Tarcher.

[lxx] Pasko Rakic, Jean-Pierre Bourgeois and Patricia S. Goldman-Rakic. "Synaptic Development of the Cerebral Cortex: Implications for Learning, Memory, and Mental Illness." *In: Progress in Brain Research* (Book Series); *In:* J. Van Pelt and others (Eds.), *The Self-Organizing Brain: From Growth Cones to Functional Networks.* (1994), Elsevier. CII, 227–243. <https://doi.org/10.1016/S0079-6123(08)60543-9>.

[lxxi] Linda Patia Spear. "Adolescent Neurodevelopment." *The Journal of Adolescent Health: Official Publication of the Society for Adolescent Medicine*, 52(202), (2013), S7–13. <https://doi.org/10.1016/j.jadohealth.2012.05.006>.

[lxxii] Takashi Kitamura and others. "Engrams and Circuits Crucial for Systems Consolidation of a Memory." *Science*, 356(6333), (2017), 73–78. (New York, N.Y.). <https://doi.org/10.1126/science.aam6808>.

[lxxiii] Elliot M. Tucker-Drob, Daniel A. Briley and K. Paige Harden. "Genetic and Environmental Influences on Cognition Across Development and Context." *Current Directions in Psychological Science*, 22(5), (2013), 349–355. <https://doi.org/10.1177/0963721413485087>.

[lxxiv] Joyce W. Lacy and Craig E.L. Stark. "The Neuroscience of Memory: Implications for the Courtroom." *Nature Reviews. Neuroscience*, 14(9), (2013), 649–658. <https://doi.org/10.1038/nrn3563>.

[lxxv] Jeffrey A. Kleim and others. "Motor Learning Induces Astrocytic Hypertrophy in the Cerebellar Cortex." *Behavioural Brain Research*, 178(2), (2007), 244–249. <https://doi.org/10.1016/j.bbr.2006.12.022>.

[lxxvi] Robert J. Zatorre, R. Douglas Fields and Heidi Johansen-Berg. "Plasticity in Gray and White: Neuroimaging Changes in Brain Structure during Learning." *Nature Neuroscience*, 15(4), (2012), 528–536. <https://doi.org/10.1038/nn.3045>.

[lxxvii] William Hirst and others. "Long-Term Memory for the Terrorist Attack of September 11: Flashbulb Memories, Event Memories, and the Factors That Influence Their Retention." *Journal of Experimental Psychology. General*, 138(2), (2009), 161–176. <https://doi.org/10.1037/a0015527>.

[lxxviii] Dr. Elizabeth Loftus and Katherine Ketcham. *The Myth of Repressed Memory: False Memories and Allegations of Sexual Abuse*. (1996). New York, NY: St. Martin's Griffin.

[lxxix] Yves Corson and Nadège Verrier. "Emotions and False Memories: Valence or Arousal?" *Psychological Science*, 18(3), (2007), 208–211. <https://doi.org/10.1111/j.1467-9280.2007.01874.x>.

[lxxx] Joseph P. Forgas, Simon M. Laham and Patrick T. Vargas. "Mood Effects on Eyewitness Memory: Affective Influences on Susceptibility to Misinformation." *Journal of Experimental Social Psychology*, 41(6), (2005), 574–588. <https://doi.org/10.1016/j.jesp.2004.11.005>.

[lxxxi] Benno Roozendaal and others. "Glucocorticoid Effects on Memory Retrieval Require Concurrent Noradrenergic Activity in the Hippocampus and Basolateral Amygdala." *The Journal of Neuroscience: The Official Journal of the Society for Neuroscience*, 24(37), (2004), 8161–8169. <https://doi.org/10.1523/JNEUROSCI.2574-04.2004>.

[lxxxii] James L. McGaugh. "Memory Consolidation and the Amygdala: A Systems Perspective." *Trends in Neurosciences*, 25(9), (2002), 456. <https://doi.org/10.1016/s0166-2236(02)02211-7>.

[lxxxiii] Erno J. Hermans and others. "How the Amygdala Affects Emotional Memory by Altering Brain Network Properties." *Neurobiology of Learning and Memory*, 112, (2014), 2–16. <https://doi.org/10.1016/j.nlm.2014.02.005>.

[lxxxiv] Michael D. Cohen and Paul Bacdayan. "Organizational Routines Are Stored As Procedural Memory: Evidence from a Laboratory Study." *Organization Science*, 5(4), (1994), 554–568.

[lxxxv] Nelson Cowan. "What Are the Differences between Long-Term, Short-Term, and Working Memory?" *Progress in Brain Research*, 169, (2008), 323–338. <https://doi.org/10.1016/S0079-6123(07)00020-9>.

[lxxxvi] K. Ashwell. *The Brain Book: Development, Function, Disorder, Health|Paperback*. (2019). Firefly Books Limited.

[lxxxvii] Eliska Prochazkova and Mariska E. Kret. "Connecting Minds and Sharing Emotions through Mimicry: A Neurocognitive Model of Emotional Contagion." *Neuroscience & Biobehavioral Reviews*, 80, (2017), 99–114. <https://doi.org/10.1016/j.neubiorev.2017.05.013>.

[lxxxviii] Endel Tulving. "Memory and Consciousness." *Canadian Psychology/Psychologie Canadienne*, 26(1), (1985), 1–12. <https://doi.org/10.1037/h0080017>.

lxxxix Konstantin Volzhenin, Jean-Pierre Changeux and Guillaume Dumas. "Multilevel Development of Cognitive Abilities in an Artificial Neural Network." *Proceedings of the National Academy of Sciences*, 119(39), (2022), e2201304119. <https://doi.org/10.1073/pnas.2201304119>.

xc Peter Gardenfors. *How Homo Became Sapiens: On the Evolution of Thinking*. Illustrated Edition (2006). Oxford; New York: Oxford University Press.

xci Patricia Curd. "Presocratic Philosophy." *Stanford Encyclopedia of Philosophy*. (2007). <https://plato.stanford.edu/ENTRIES/presocratics/> [accessed 17 February 2024].

xcii Philip Johnson-Laird. *Mental Models* (1983). Cambridge, Mass: Harvard University Press; P.N. Johnson-Laird, Sangeet S. Khemlani and Geoffrey P. Goodwin. "Logic, Probability, and Human Reasoning." *Trends in Cognitive Sciences*, 19(4), (2015), 201–214. <https://doi.org/10.1016/j.tics.2015.02.006>.

xciii Judea Pearl. *Causality*. 2nd Edition (2009). Cambridge New York, NY Port, Melbourne, New Delhi, Singapore: Cambridge University Press,

xciv Alan M. Turing. "Computing Machinery and Intelligence." *Mind*, 59, (October 1950), 433–460. <https://doi.org/10.1093/mind/lix.236.433>.

xcv Yihan Cao and others. "A Comprehensive Survey of AI-Generated Content (AIGC): A History of Generative AI from GAN to ChatGPT." (2023), *arXiv*, <https://doi.org/10.48550/arXiv.2303.04226>.

xcvi Priyantha Wijayatunga. "Probability, Paradoxes and Human Thinking" (presented at the 15th Conference of the Swedish Cognitive Science Society (SweCog 2019), 7–8 November, 2019, Umeå, Sweden, University of Skövde (2019), 54–56. <https://urn.kb.se/resolve?urn=urn:nbn:se:umu:diva-165061> [accessed 17 February 2024].

xcvii Pierre Simon Marquis De Laplace. *A Philosophical Essay on Probabilities*. Trans. by F.W. Truscott and F.L. Emory (2007). Cosimo Classics.

xcviii Stephen Herrero and others. "Fatal Attacks by American Black Bear on People: 1900–2009." *The Journal of Wildlife Management*, 75(3), (2011), 596–603. <https://doi.org/10.1002/jwmg.72>.

xcvix Richard C. Burgess. "Evaluation of Brain Connectivity: The Role of Magnetoencephalography." *Epilepsia*, 52(s4), (2011), 28–31. <https://doi.org/10.1111/j.1528-1167.2011.03148.x>.

c Christos Papadelis and others. "Can Magnetoencephalography Track the Afferent Information Flow along White Matter Thalamo-Cortical Fibers?" *NeuroImage*, 60(2), (2012), 1092–1105. <https://doi.org/10.1016/j.neuroimage.2012.01.054>.

ci Earl Hunt. *Human Intelligence*. Illustrated Edition. (2010). Cambridge University Press.

cii Anna T. Cianciolo and Robert J. Sternberg. *Intelligence: A Brief History*. (2004). Blackwell Publishing.

ciii Robert J. Sternberg and Elena L. Grigorenko. "Intelligence and Culture: How

Culture Shapes What Intelligence Means, and the Implications for a Science of Well-Being." *In*: Felicia A. Huppert, Nick Baylis and Barry Keverne (Eds.), *The Science of Well-Being*. (2005). Oxford University Press. <https://doi.org/10.1093/acprof:oso/9780198567523.003.0014>.

civ Richard J. Haier. *The Neuroscience of Intelligence*. (2016). New York, NY: Cambridge University Press.

cv Arija G. Jansen and others. "What Twin Studies Tell Us About the Heritability of Brain Development, Morphology, and Function: A Review." *Neuropsychology Review*, 25(1), (2015), 27–46. <https://doi.org/10.1007/s11065-015-9278-9>.

cvi Ron Oostdam and Joost Meijer. "Influence of Test Anxiety on Measurement of Intelligence." *Psychological Reports*, 92(1), (2003), 3–20. <https://doi.org/10.2466/pr0.2003.92.1.3>.

cvii Robert J. Sternberg. *Beyond IQ: A Triarchic Theory of Human Intelligence*. (1985). CUP Archive.

cviii Brian A. Wandell, Serge O. Dumoulin and Alyssa A. Brewer. "Visual Cortex in Humans." *Encyclopedia of Neuroscience*, 10, (2009), 251–257.

cix Giulia Enders. *Gut: The Inside Story of Our Body's Most Underrated Organ*. Revised Edition. (2018). Greystone Books.

cx Michael Schemann, Thomas Frieling and Paul Enck. "To Learn, to Remember, to Forget: How Smart Is the Gut?" *Acta Physiologica*, 228(1), (2020), e13296. Oxford, England. <https://doi.org/10.1111/apha.13296>.

cxi Matteo Briguglio and others. "Dietary Neurotransmitters: A Narrative Review on Current Knowledge." *Nutrients*, 10(5), (2018), 591. <https://doi.org/10.3390/nu10050591>.

cxii Masanao Toda. "Emotion and Decision Making." *Acta Psychologica*, 45(1), (1980), 133–155. <https://doi.org/10.1016/0001-6918(80)90026-8>.

cxiii Robert J. Sternberg, Henry L. Roediger and Diane F. Halpern. *Critical Thinking in Psychology*. (2007). Cambridge University Press.

cxiv James Wood. "Chapter 3 – Radiation Sources." *In*: James Wood (Ed.), *Computational Methods in Reactor Shielding*. (1982), 56–78. Pergamon. <https://doi.org/10.1016/B978-0-08-028685-3.50006-X>.

cxv Jingjie Yeo and others. "Multiscale Modeling of Keratin, Collagen, Elastin and Related Human Diseases: Perspectives from Atomistic to Coarse-Grained Molecular Dynamics Simulations." *Extreme Mechanics Letters*, 20, (2018), 112–124. <https://doi.org/10.1016/j.eml.2018.01.009>.

cxvi Justin J. Patricia and Amit S. Dhamoon. "Physiology, Digestion." *In*: *StatPearls*. (2024). Treasure Island (FL): StatPearls Publishing. <http://www.ncbi.nlm.nih.gov/books/NBK544242/> [accessed 20 February 2024].

cxvii C. Allen and L. Manson. "7 – Managing Medical Radioisotope Production Facilities." *In*: Jas Devgun (Ed.), *Managing Nuclear Projects*. (2013), 136–151. Woodhead Publishing Series in Energy, Woodhead Publishing. <https://doi.org/10.1533/9780857097262.2.136>.

cxviii Maria A. Ermolaeva, Alexander Dakhovnik and Björn Schumacher. "Quality Control Mechanisms in Cellular and Systemic DNA Damage Responses." *Ageing Research Reviews, Quality Control Systems in Aging*, 23, (2015), 3–11. <https://doi.org/10.1016/j.arr.2014.12.009>.

cxix Edward O. Wilson. *Genesis: The Deep Origin of Societies*. Illustrated Edition. (2019). New York: Liveright.

cxx Rashelle Ripa and others. "Physiology, Cardiac Muscle." *In: StatPearls*. (2024). Treasure Island (FL): StatPearls Publishing. <http://www.ncbi.nlm. nih.gov/books/NBK572070/> [accessed 20 February 2024].

cxxi C. Tommaso, N. Belic and M. Brandfonbrener. "Asynchronous Ventricular Pacing: A Rare Cause of Ventricular Tachycardia." *Pacing and Clinical Electrophysiology: PACE*, 5(4), (1982), 561–563. <https:// doi.org/10.1111/j.1540-8159.1982.tb02278.x>.{\\i{}Pacing and Clinical Electrophysiology: PACE}, 5.4 (1982).

cxxii Antonis A. Manolis and others. "The Role of the Autonomic Nervous System in Cardiac Arrhythmias: The Neuro-Cardiac Axis, More Foe than Friend?" *Trends in Cardiovascular Medicine*, 31(5), (2021), 290–302. <https://doi.org/10.1016/j.tcm.2020.04.011>.

cxxiii Simon Baron-Cohen. *The Pattern Seekers: How Autism Drives Human Invention*. (2020). New York (N.Y.): Basic Books.

cxxiv Rudolf Carnap. *Philosophical Foundations of Physics: An Introduction to the Philosophy of Science*. https://www.amazon.com>Philosophical-Foundations-Physics-Introduction-Philosophy/dpB0000CN9NI.

cxxv Yuval Noah Harari. *Sapiens: A Brief History of Humankind*. (2016). Signal.

cxxvi UN Climate Change Conference – United Arab Emirates | UNFCCC. <https://unfccc.int/cop28> [accessed 17 February 2024].

cxxvii Edward N. Lorenz. *The Essence of Chaos*. 1st Edition. (1995). Seattle: University of Washington Press.

cxxviii Steven H. Strogatz. *Nonlinear Dynamics And Chaos: With Applications To Physics, Biology, Chemistry, And Engineering*. 1st Edition. (2001). Cambridge, Mass: Westview Press.

cxxix Hansulrich Steimle and Carol Norberg. "Astronaut Selection and Training. *In*: Carol Norberg (Ed.), *Human Spaceflight and Exploration*. Springer Praxis Books. (2013), 255–294. Berlin, Heidelberg: Springer. <https://doi. org/10.1007/978-3-642-23725-6_7>.

cxxx Châu Bao Ngô - Scholars | Institute for Advanced Study, 2019. <https:// www.ias.edu/scholars/bao-ch%C3%A2u-ng%C3%B4> [accessed 17 February 2024].

cxxxi Michel Foucault. *Power: Essential Works of Foucault, 1954-1984*. Edited by James D. Faubion, translated by Robert Hurley. (2001). New York: The New Press.

cxxxii Davide Gnocchi and Giovannella Bruscalupi. "Circadian Rhythms and Hormonal Homeostasis: Pathophysiological Implications." *Biology*, 6(1), (2017), 10. <https://doi.org/10.3390/biology6010010>; Martin E. Young.

"The Cardiac Circadian Clock." *JACC: Basic to Translational Science*, 8(12), (2023), 1613–1628. <https://doi.org/10.1016/j.jacbts.2023.03.024>; Tatiana M. Anderson and Jan-Marino Ramirez. "Respiratory Rhythm Generation: Triple Oscillator Hypothesis." *F1000Research*, (2017). <https://doi.org/10.12688/f1000research.10193.1>.

[cxxxiii] Joachim D. Pleil and others. "The Physics of Human Breathing: Flow, Timing, Volume, and Pressure Parameters for Normal, on-Demand, and Ventilator Respiration." *Journal of Breath Research*, 15(4), (2021), 10.1088/1752-7163/ac2589. <https://doi.org/10.1088/1752-7163/ac2589>.

[cxxxiv] Rümeysa İnce, Saliha Seda Adanır, and Fatma Sevmez. "The Inventor of Electroencephalography (EEG): Hans Berger (1873–1941)." *Child's Nervous System*, 37(9), (2021), 2723–2724. <https://doi.org/10.1007/s00381-020-04564-z>.

[cxxxv] Juri D. Kropotov. "Chapter 2.3: Beta and Gamma Rhythms." *In*: Juri D. Kropotov (Ed.), *Functional Neuromarkers for Psychiatry*. (2016), 107–119. San Diego: Academic Press. <https://doi.org/10.1016/B978-0-12-410513-3.00009-7>.

[cxxxvi] Sergey Nurk and others. "The Complete Sequence of a Human Genome." *Science*, 376(6588), (2022), 44–53. <https://doi.org/10.1126/science.abj6987>.

[cxxxvii] Leroy Hood and Lee Rowen. "The Human Genome Project: Big Science Transforms Biology and Medicine." *Genome Medicine*, 5(9) (2013), 79. <https://doi.org/10.1186/gm483>.

[cxxxviii] Miryam Naddaf. "Europe Spent €600 million to Recreate the Human Brain in a Computer. How Did It Go?" *Nature*, 620(7975), (2023), 718–720. <https://doi.org/10.1038/d41586-023-02600-x>.

[cxxxix] Katrin Amunts and others. *The Coming Decade of Digital Brain Research: A Vision for Neuroscience at the Intersection of Technology and Computing*. (23 March 2023). Zenodo. <https://doi.org/10.5281/zenodo.7764003>.

[cxl] Per Bak. *How Nature Works: The Science of Self-Organized Criticality*. 1996th Edition (New York, NY, USA: Springer/Sci-Tech/Trade, 2003).

[cxli] Edward N. Lorenz. "Deterministic Nonperiodic Flow. *Journal of the Atmospheric Sciences*, 2(2), (1963), 130–141. <https://doi.org/10.1175/1520-0469(1963)020<0130:DNF>2.0.CO;2>.

[cxlii] John Jumper and others. "Highly Accurate Protein Structure Prediction with AlphaFold." *Nature*, 596(7873), (2021), 583–589. <https://doi.org/10.1038/s41586-021-03819-2>.

[cxliii] Arkady Pikovsky, Michael Rosenblum and Jürgen Kurths. *Synchronization: A Universal Concept in Nonlinear Sciences*. Illustrated Edition. (2003). Cambridge: Cambridge University Press.

[cxliv] Bicky A. Márquez, José J. Suárez-Vargas and Javier A. Ramírez. "Polynomial Law for Controlling the Generation of N-Scroll Chaotic Attractors in an Optoelectronic Delayed Oscillator." *Chaos: An Interdisciplinary Journal of Nonlinear Science*, 24(3), (2014), 033123. <https://doi.org/10.1063/1.4892947>.

[cxlv] S.H. Strogatz. *Nonlinear Dynamics and Chaos: With Applications to Physics, Biology, Chemistry, and Engineering.* (2001). Cambridge, Mass.: Westview Press.

[cxlvi] Holger Kantz and Thomas Schreiber. *Nonlinear Time Series Analysis.* 2nd Edition. (2003). Cambridge, UK; New York: Cambridge University Press.

[cxlvii] A. Pikovsky, M. Rosenblum and J. Kurths. *Synchronization: A Universal Concept in Nonlinear Sciences.* (2003). Cambridge: Cambridge University Press.

[cxlviii] Hartmut Rosa. *Resonance: A Sociology of Our Relationship to the World.* Translated by James Wagner, 1st Edition. (2019). Medford, MA: Polity.

[cxlix] Thomas E. Murphy and others. "Complex Dynamics and Synchronization of Delayed-Feedback Nonlinear Oscillators." *Philosophical Transactions: Mathematical, Physical and Engineering Sciences*, 368(1911), (2010), 343–366.

[cl] Juergen Fell and Nikolai Axmacher. "The Role of Phase Synchronization in Memory Processes." *Nature Reviews Neuroscience*, 12(2), (2011), 105–118. <https://doi.org/10.1038/nrn2979>.

[cli] Robert M. French and Elizabeth Thomas. "The Dynamical Hypothesis in Cognitive Science: A Review Essay of Mind As Motion." *Minds and Machines*, 11(1), (2001), 101–111. <https://doi.org/10.1023/A:1011256824648>.

[clii] Alejandro Pérez, Manuel Carreiras and Jon Andoni Duñabeitia. "Brain-to-Brain Entrainment: EEG Interbrain Synchronization While Speaking and Listening." *Scientific Reports*, 7(1), (2017), 4190. <https://doi.org/10.1038/s41598-017-04464-4>.

[cliii] Evan G. Antzoulatos and Earl K. Miller. "Increases in Functional Connectivity between Prefrontal Cortex and Striatum during Category Learning." *Neuron*, 83(1), (2014), 216–225. <https://doi.org/10.1016/j.neuron.2014.05.005>.

[cliv] Bernard Widrow and Michael A. Lehr. "Artificial Neural Networks of the Perceptron, Madaline, and Backpropagation Family." *In*: Hans-Werner Bothe, Madjid Samii and Rolf Eckmiller (Eds.), *Neurobionics.* (1993), 133–205. Amsterdam: Elsevier. <https://doi.org/10.1016/B978-0-444-89958-3.50013-9>.

[clv] Daniel Redhead and Eleanor A. Power. "Social Hierarchies and Social Networks in Humans." *Philosophical Transactions of the Royal Society B: Biological Sciences*, 377(1845), (2022), 20200440. <https://doi.org/10.1098/rstb.2020.0440>.

[clvi] Fawaz A. Gerges. "ISIS and the Third Wave of Jihadism." *Current History*, 113(767), (2014), 339–343.

[clvii] Andrew M. Pettigrew. "Information Control as a Power Resource." *Sociology*, 6(2), (1972), 187–204. <https://doi.org/10.1177/003803857200600202>.

clviii Aminia M. Brueggemann and Peter Schulman. *Rhine Crossings: France and Germany in Love and War.* (2006). State University of New York Press.

clix Agust Arnorsson and Gylfi Zoega. "On the Causes of Brexit." *European Journal of Political Economy*, 55, (2018), 301–323. <https://doi.org/10.1016/j.ejpoleco.2018.02.001>.

clx Scott Kuindersma and others. "Optimization-Based Locomotion Planning, Estimation, and Control Design for the Atlas Humanoid Robot." *Autonomous Robots*, 40(3), (2016), 429–455. <https://doi.org/10.1007/s10514-015-9479-3>.

clxi L. Dartnell. *Origins: How Earth's History Shaped Human History.* (2019). New York: Basic Books.

clxii Simson L. Garfinkel and Rachel H. Grunspan. *The Computer Book: From the Abacus to Artificial Intelligence, 250 Milestones in the History of Computer Science.* Illustrated Edition. (2018). New York: Union Square & Co.

clxiii John McCarthy and others. "A Proposal for the Dartmouth Summer Research Project on Artificial Intelligence." *AI Magazine*, 27(4), (1955), 12–12. <https://doi.org/10.1609/aimag.v27i4.1904>; Turing.

clxiv Yihan Cao and others. A Comprehensive Survey of AI-Generated Content (AIGC): A History of Generative AI from GAN to ChatGPT. (2023) Preprint at https://doi.org/10.48550/arXiv.2303.04226.

clxv Timothy P. Lillicrap and others. "Random Synaptic Feedback Weights Support Error Backpropagation for Deep Learning." *Nature Communications*, 7(1), (2016), 13276. <https://doi.org/10.1038/ncomms13276>.

clxvi T.P. Lillicrap, Cownden, D., Tweed, D.B. and Akerman, C.J. "Random Synaptic Feedback Weights Support Error Backpropagation for Deep Learning." *Nat Commun*, 7, (2016), 13276.

clxvii Benjamin Scellier and Yoshua Bengio "Equilibrium Propagation: Bridging the Gap between Energy-Based Models and Backpropagation." *Frontiers in Computational Neuroscience*, 11, (2017). <https://www.frontiersin.org/articles/10.3389/fncom.2017.00024> [accessed 17 February 2024].

clxviii B. Widrow and M.A. Lehr. Artificial Neural Networks of the Perceptron, Madaline, and Backpropagation Family. *In:* Bothe, H.-W., Samii, M. & Eckmiller, R. (eds.), Neurobionics. 133–205 (1993). Elsevier, Amsterdam. doi:10.1016/B978-0-444-89958-3.50013-9.

clxix Murali Emani and others. "A Comprehensive Evaluation of Novel AI Accelerators for Deep Learning Workloads." *In*: 2022 IEEE/ACM International Workshop on Performance Modeling, Benchmarking and Simulation of High Performance Computer Systems (PMBS), 2022, 13–25, <https://doi.org/10.1109/PMBS56514.2022.00007>; Albert Reuther and others. "Survey of Machine Learning Accelerators." *In*: 2020 IEEE High Performance Extreme Computing Conference (HPEC), 2020, 1–12. <https://doi.org/10.1109/HPEC43674.2020.9286149>.

clxx Wei Cao and others. "The Future Transistors." *Nature*, 620(7974), (2023), 501–515. <https://doi.org/10.1038/s41586-023-06145-x>.

clxxi Ian Goodfellow, Yoshua Bengio and Aaron Courville. *Deep Learning.* (2016). Cambridge, Massachusetts: The MIT Press.

clxxii Alec Radford and others "Language Models are Unsupervised Multitask Learners." (2019). <https://www.semanticscholar.org/paper/Language-Models-are-Unsupervised-Multitask-Learners-Radford-Wu/9405cc0d616 9988371b2755e573cc28650d14dfe> [accessed 17 February 2024].

clxxiii Radford, A. et al. "Language Models are Unsupervised Multitask Learners." (2019). <https://www.semanticscholar.org/paper/Language-Models-are-Unsupervised-Multitask-Learners-Radford-Wu/9405cc0d6169988371b27 55e573cc28650d14dfe>

clxxiv Helene Darius, *Savant Syndrome – Theories and Empirical Findings.* (2007). Institutionen för kommunikation och information. <https://urn.kb.se/resolve?urn=urn:nbn:se:his:diva-52> [accessed 17 February 2024].

clxxv Sven Bölte and Fritz Poustka. "Comparing the Intelligence Profiles of Savant and Nonsavant Individuals with Autistic Disorder." *Intelligence*, 32(2), (2004), 121–131. <https://doi.org/10.1016/j.intell.2003.11.002>.

clxxvi Scott Barry Kaufman, *Ungifted: Intelligence Redefined.* 1st Edition. (2013). New York: Basic Books.

clxxvii Darold A. Treffert. "The Savant Syndrome: An Extraordinary Condition. A Synopsis: Past, Present, Future." *Philosophical Transactions of the Royal Society B: Biological Sciences*, 364(1522), (2009), 1351–1357. <https://doi.org/10.1098/rstb.2008.0326>.

Index